New

WITHDRAWN

SCHAUM'S *Easy* OUTLINES

Other Books in Schaum's Easy Outline Series include:

Schaum's Easy Outline: Calculus
Schaum's Easy Outline: College Physics
Schaum's Easy Outline: Statistics
Schaum's Easy Outline: Programming with C++
Schaum's Easy Outline: College Chemistry
Schaum's Easy Outline: French
Schaum's Easy Outline: Spanish
Schaum's Easy Outline: German
Schaum's Easy Outline: Organic Chemistry

SCHAUM'S *Easy* OUTLINES

COLLEGE
ALGEBRA

BASED ON SCHAUM'S
Outline of College Algebra
BY MURRAY R. SPIEGEL AND
ROBERT E. MOYER

ABRIDGEMENT EDITOR:
GEORGE J. HADEMENOS

SCHAUM'S OUTLINE SERIES
McGRAW-HILL

*New York San Francisco Washington, D.C. Auckland Bogotá
Caracas Lisbon London Madrid Mexico City Milan Montreal
New Delhi San Juan Singapore Sydney Tokyo Toronto*

MURRAY R. SPIEGEL received the M.S. degree in Physics and the Ph.D. in Mathematics from Cornell University. His last position was Professor and Chairman of Mathematics at the Rensselaer Polytechnic Institute, Hartford Graduate Center.

ROBERT E. MOYER has been teaching mathematics at Fort Valley State University in Fort Valley, Georgia, since 1985. He received his Doctor of Philosophy in Mathematics Education from the University of Illinois in 1974 and his Master of Science in 1967 and his Bachelor of Science in 1964 from Southern Illinois University.

GEORGE J. HADEMENOS has taught at the University of Dallas and has performed research at the University of California at Los Angeles and the University of Massachusettes Medical Center. He earned a Bachelor of Science in Physics degree from Angelo State University and the Master of Science and Doctor of Philosophy in Physics from the University of Texas at Dallas.

3 4 5 6 7 8 9 10 11 12 13 14 15 DOC DOC 9 0 9 8 7 6 5 4 3 2 1

ISBN 0-07-052709-1

Sponsoring Editor: Barbara Gilson
Production Supervisor: Tina Cameron
Editing Supervisor: Maureen B. Walker

McGraw-Hill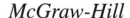

A Division of The McGraw-Hill Companies

Contents

Chapter 1   Fundamental Tools of Algebra   1

Chapter 2   Algebraic Expressions
             and Operations   18

Chapter 3   Functions   38

Chapter 4   Linear Equations   70

Chapter 5   Quadratic Equations   100

Chapter 6   Sequences, Series, and
             Mathematical Induction   110

Chapter 7   Permutations, Combinations,
             the Binomial Theorem,
             and Probability   118

Index   132

Chapter 1

FUNDAMENTAL TOOLS OF ALGEBRA

IN THIS CHAPTER:

✔ *Fundamental Operations with Numbers*
Properties of Real Numbers
Exponents and Powers
Logarithms
Radicals
Complex Numbers

Fundamental Operations with Numbers

Four Operations with Numbers

The four fundamental operations in algebra are

1

- **Addition:** When two numbers a and b are added, their sum is indicated by $a + b$. Thus, $3 + 2 = 5$.

- **Subtraction:** When a number b is subtracted from a number a, the difference is indicated by $a - b$. Thus, $6 - 2 = 4$.

- **Multiplication:** The product of two numbers a and b is a number c such that $a \cdot b = c$. The operation of multiplication may be indicated by a cross, a dot, or parentheses. Thus, $5 \times 3 = 5 \cdot 3 = (5)(3) = 15$.

- **Division:** When a number a is divided by a number b, the quotient obtained is written

  $$a \div b \text{ or } \frac{a}{b} \text{ or } a/b$$

  where a is the dividend and b the divisor. The expression a/b is also called a fraction, having numerator a and denominator b.
  Division by zero is not defined.

System of Real Numbers

The system of real numbers includes the following:

- *Natural numbers*, written 1, 2, 3, 4, . . . , used in counting are also known as positive integers. If two such numbers are added or multiplied, the result is always a natural number.

- *Positive rational numbers* or positive fractions are the quotients of two positive integers, such as 2/3, 8/5, 121/17. The positive rational numbers include the set of natural numbers. Thus, the rational number 3/1 is the natural number 3.

- *Positive irrational numbers* are numbers that are not rational, that is, that cannot be written as quotients of two integers such as $\sqrt{2}$ and π.

- *Zero*, written 0, enlarged the number system to permit such operations such as 6 -6 or 10 -10. Zero has the property that any number multiplied by zero is zero. Zero divided by any number that is not equal to zero is zero.

- *Negative integers, negative rational numbers, and negative irrational numbers* such as -3, -2/3, and $-\sqrt{2}$, enlarged the number system to permit such operations as 2 - 8, π - 3π, or 2 -2$\sqrt{2}$.

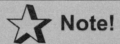 **Note!**

The real number system consists of the collection of positive and negative rational and irrational numbers and zero.

Graphical Representation of Real Numbers

It is often useful to represent real numbers by points on a line. To do this, we choose a point on the line to represent the real number zero and call this point the origin. The positive integers +1, +2, +3, . . . are then associated with points on the line at distances 1, 2, 3, . . . units, respectively, to the *right* of the origin (see Figure 1-1), while the negative integers -1, -2, -3, . . . are associated with points on the line at distances 1, 2, 3, . . . units respectively to the *left* of the origin.

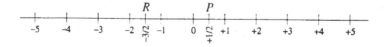

Figure 1-1

The rational number 1/2 is represented on this scale by a point P halfway between 0 and +1. The negative number -3/2 or -1 1/2 is represented by a point R 1 1/2 units to the left of the origin.

You Need to Know

The position of real numbers on a line establishes an order to the real number system. If a point A lies to the right of another point B on the line, we say that the number corresponding to A is *greater* or *larger* than the number corresponding to B, or that the number corresponding to B is *less* or *smaller* than the number corresponding to A.

Sets of Real Numbers

The system of real numbers can be expressed in terms of sets. A set

- Has closure under an operation if the result of performing the operation with two elements of the set is also an element of the set. The set X is closed under the operation * if, for all elements a and b in set X, the result $a*b$ is in set X.

- Has an identity under an operation if there is an element in the set that, when combined with each element in the set, leaves that element unchanged. The set X has an identity under the operation * if there is an element j in set X such that $j*a = a*j = a$ for all elements a in set X.

- Has inverses under an operation if for each element of the set there is an element of the set such that when these two elements are combined using the operation, the result is the identity for the set under the operation. If a set does not have an identity under an operation, it cannot have the inverse property for the operation. If X is a set that has identity j under operation, then it has inverses if for each element a in set X there is an element a' in set X such that $a*a' = j$ and $a'*a = j$.

- Under an operation may also have the associative property and the commutative property. If there are two operations on the set, then the set could have the distributive property.

Properties of Real Numbers

- *Commutative property for addition.* The order of addition of two numbers does not affect the result. Thus,

  $a + b = b + a \qquad 5 + 3 = 3 + 5 = 8$

- *Associative property for addition.* The terms of a sum may be grouped in any manner without affecting the result.

  $a + (b + c) = (a + b) + c = a + b + c$

  $3 + (4 + 1) = (3 + 4) + 1 = 3 + 4 + 1 = 8$

- *Commutative property for multiplication.* The order of the factors of a product does not affect the result.

  $a \cdot b = b \cdot a \qquad 2 \cdot 5 = 5 \cdot 2 = 10$

- *Associative property for multiplication.* The factors of a product may be grouped in any manner without affecting the result.

  $a(bc) = (ab)c = abc \qquad 3(4 \cdot 6) = (3 \cdot 4)6 = 3 \cdot 4 \cdot 6 = 72$

- *Distributive property for multiplication over addition.* The product of a number a by the sum of two numbers $(b + c)$ is equal to the sum of the products ab and ac.

$$a(b + c) = ab + ac, \qquad 4(3 + 2) = 4 \cdot 3 + 4 \cdot 2 = 20$$

Example 1.1 Which properties are true for the counting numbers, whole numbers, integers, rational numbers, irrational numbers, and real numbers under the operation of addition?

+	Counting	Whole	Integers	Rational	Irrational	Real
Closure	Yes	Yes	Yes	Yes	No	Yes
Identity	No	Yes	Yes	Yes	No	Yes
Inverse	No	No	Yes	Yes	No	Yes
Associativity	Yes	Yes	Yes	Yes	Yes	Yes
Commutativity	Yes	Yes	Yes	Yes	Yes	Yes

There are some properties that sets of numbers have that do not depend on an operation to be true. Three such properties are **order, density**, and **completeness**.

A set of numbers has *order* if, given two distinct elements in the set, one element is greater than the other.

A set of numbers has *density* if, between any two elements of the set, there is another element of the set.

A set of numbers has *completeness* if the points using its elements as coordinates completely fill a line or plane.

Example 1.2 Which properties are true for the counting numbers, whole numbers, integers, rational numbers, irrational numbers, and real numbers?

	Counting	Whole	Integers	Rational	Irrational	Real
Order	Yes	Yes	Yes	Yes	Yes	Yes
Density	No	No	No	Yes	Yes	Yes
Completeness	No	No	No	No	No	Yes

Rules of Signs

- To add two numbers with like signs, add their absolute values and prefix the common sign. The absolute value of a real number, a, is defined as the distance in units from the point whose coordinate is a to the origin. Thus,

Examples 1.3 $\qquad$ $3 + 4 = 7;$ $\qquad$ $(-3) + (-4) = -7$

- To add two numbers with unlike signs, find the difference between their absolute values and prefix the sign of the number with greater absolute value.

Examples 1.4 $\qquad$ $17 + (-8) = 9;$ $\qquad$ $(-6) + 4 = -2$

- To subtract one number b from another number a, change the operation to addition and replace b by its opposite, $-b$.

Examples 1.5 $\qquad$ $12 - (7) = 12 + (-7) = 5$
$\qquad\qquad\qquad$ $(-9) - (4) = -9 + (-4) = -13$

- To multiply (or divide) two numbers having like signs, multiply (or divide) their absolute values and prefix a plus sign (or no sign).

Examples 1.6 $\qquad$ $(5)(3) = 15;$ $\qquad$ $(-5)(-3) = 15$
$\qquad\qquad\qquad$ $-6/-3 = 2$

- To multiply (or divide) two numbers having unlike signs, multiply (or divide) their absolute values and prefix a minus sign.

Examples 1.7 $(-3)(6) = -18$ $(3)(-6) = -18$
$-12/4 = -3$

Operations with Fractions

Operations with fractions may be performed according to the following rules:

- The value of a fraction remains the same if its numerator and denominator are both multiplied or divided by the same number provided the number is not zero.

Examples 1.8

$$\frac{3}{4} = \frac{3 \cdot 2}{4 \cdot 2} = \frac{6}{8}, \quad \frac{15}{18} = \frac{15 \div 3}{18 \div 3} = \frac{5}{6}$$

- Changing the sign of the numerator or denominator of a fraction changes the sign of the fraction.

Example 1.9

$$\frac{-3}{5} = -\frac{3}{5} = \frac{3}{-5}$$

- Adding two fractions with a common denominator yields a fraction whose numerator is the sum of the numerators of the given fractions and whose denominators is the common denominator.

Example 1.10

$$\frac{3}{5} + \frac{4}{5} = \frac{3+4}{5} = \frac{7}{5}$$

- The sum or difference of two fractions having different denominators may be found by writing the fractions with a common denominator.

Example 1.11

$$\frac{1}{4} + \frac{2}{3} = \frac{3}{12} + \frac{8}{12} = \frac{11}{12}$$

- The product of two fractions is a fraction whose numerator is the product of the numerators of the given fractions and whose denominator is the product of the denominators of the fractions.

Examples 1.12

$$\frac{2}{3} \cdot \frac{4}{5} = \frac{2 \cdot 4}{3 \cdot 5} = \frac{8}{15}, \quad \frac{3}{4} \cdot \frac{8}{9} = \frac{3 \cdot 8}{4 \cdot 9} = \frac{24}{36} = \frac{2}{3}$$

- The reciprocal of a fraction is a fraction whose numerator is the denominator of the given fraction and whose denominator is the numerator of the given fraction. Thus, the reciprocal of 3 (i.e., 3/1) is 1/3. Similarly, the reciprocals of 5/8 and -4/3 are 8/5 and 3/-4 or -3/4, respectively.

- To divide two fractions, multiply the first by the reciprocal of the second.

Examples 1.13

$$\frac{a}{b} \div \frac{c}{d} = \frac{a \cdot d}{b \cdot c} = \frac{ad}{bc}, \quad \frac{2}{3} \div \frac{4}{5} = \frac{2 \cdot 5}{3 \cdot 4} = \frac{10}{12} = \frac{5}{6}$$

Exponents and Powers

When a number a is multiplied by itself n times, the product $a \cdot a \cdot a \cdots a$ (n times) is indicated by the symbol a^n which is referred to as "the nth power of a" or "a to the nth power" or "a to the nth." In a^n, the number a is called the *base* and the positive integer n is the *exponent*.

Examples 1.14

$2 \cdot 2 \cdot 2 \cdot 2 \cdot 2 = 2^5 = 32$

$(-5)^3 = (-5)(-5)(-5) = -125$

$2 \cdot x \cdot x \cdot x = 2x^3$

$a \cdot a \cdot a \cdot b \cdot b = a^3 b^2$

$(a - b)(a - b)(a - b) = (a - b)^3$

If p and q are positive integers, then the following are laws of exponents:

(1) $a^p \cdot a^q = a^{p+q}$

$2^3 \cdot 2^4 = 2^{3+4} = 2^7 = 128$

(2) $a^p/a^q = a^{p-q} = 1/a^{q-p}$ if $a \neq 0$

$3^5/3^2 = 3^{5-2} = 3^3, \; 3^4/3^6 = 1/3^{6-4} = 1/3^2$

(3) $(a^p)^q = a^{pq}$

$(4^2)^3 = 4^6, \quad\quad (3^4)^2 = 3^8$

(4) $(ab)^p = a^p b^p, \; (a/b)^p = a^p/b^p$ if $b \neq 0$

$(4 \cdot 5)^2 = 4^2 \cdot 5^2, \; (5/2)^3 = 5^3/2^3$

(5) $a^{-p} = 1/a^p$ if $a \neq 0$

$2^{-4} = 1/2^4 = 1/16, \; 1/3^{-3} = 3^3 = 27, \quad -4x^{-2} = -4/x^2,$

$(a + b)^{-1} = 1/(a + b)$

Logarithms

If $b^x = N$, where N is a positive number and b is a positive number different from 1, then the exponent x is the logarithm of N to the base b and is written $x = \log_b N$.

Example 1.15 Write $3^2 = 9$ using logarithmic notation.

Since $3^2 = 9$, then 2 is the logarithm of 9 to the base 3 or $2 = \log_3 9$.

Example 1.16 Evaluate $\log_2 8$.

$\log_2 8$ is that number x to which the base ② must be raised in order to yield 8 or $2^x = 8$, x = 3. Hence, $\log_2 8 = 3$.

Both $b^x = N$ and $x = \log_b N$ are equivalent relationships; $b^x = N$ is called the exponential form and $x = \log_b N$ the logarithmic form of the relationship. As a consequence, corresponding to laws of exponents there are laws of logarithms.

Laws of Logarithms

I. The logarithm of the product of two positive numbers M and N is equal to the sum of the logarithms of the numbers, i.e.,

$$\log_b MN = \log_b M + \log_b N$$

Example 1.17 Express $\log_2 3(5)$ in terms of simpler logarithms.

$$\log_2 3(5) = \log_2 3 + \log_2 5$$

II. The logarithm of the quotient of two positive numbers M and N is equal to the difference of the logarithms of the numbers, i.e.,

$$\log_b(M/N) = \log_b M - \log_b N$$

Example 1.18 Express $\log_{10} (17/24)$ in terms of simpler logarithms.

$$\log_{10} (17/24) = \log_{10} 17 - \log_{10} 24$$

III. The logarithm of the pth power of a positive number M is equal to p multiplied by the logarithm of the number, i.e.,

$$\log_b M^p = p \log_b M$$

Example 1.19 Evaluate $\log_7 5^3$.

$$\log_7 5^3 = 3 \log_7 5$$

Remember!

Natural Logarithms

The system of logarithms whose base is the constant e is called the natural logarithm system. The number e is an irrational number that is defined as e = 2.718281828. The base of a logarithm in e is written as ln. The exponential form of ln a = b is $e^b = a$. Thus, ln 25 = $\log_e$ 25.

Radicals

A radical is an expression of the form $\sqrt[n]{a}$ which denotes the principal nth root of a. The positive integer n is the index, or order, of the radical and the number a is the radicand. The index is omitted if $n = 2$.

Laws of Radicals

Writing $\sqrt[n]{a} = a^{\frac{1}{n}}$ makes the laws of radicals the same as the laws for exponents. The following are the laws most frequently used. Note: If n is even, assume a, b $\geq$ 0.

(1) $\left(\sqrt[n]{a}\right)^n = a$

Examples 1.20

$$(\sqrt[3]{6})^3 = 6, \qquad (\sqrt[4]{x^2 + y^2})^4 = x^2 + y^2$$

(2) $\sqrt[n]{ab} = \sqrt[n]{a} \, \sqrt[n]{b}$

Examples 1.21

$$\sqrt[3]{54} = \sqrt[3]{27 \cdot 2} = \sqrt[3]{27} \cdot \sqrt[3]{2} = 3\sqrt[3]{2}, \qquad \sqrt[7]{x^2 y^5} = \sqrt[7]{x^2} \, \sqrt[7]{y^5}$$

(3) $\sqrt[n]{\dfrac{a}{b}} = \dfrac{\sqrt[n]{a}}{\sqrt[n]{b}} \qquad b \neq 0$

Examples 1.22

$$\sqrt[5]{\frac{5}{32}} = \frac{\sqrt[5]{5}}{\sqrt[5]{32}} = \frac{\sqrt[5]{5}}{2}, \qquad \sqrt[3]{\frac{(x+1)^3}{(y-2)^6}} = \frac{\sqrt[3]{(x+1)^3}}{\sqrt[3]{(y-2)^6}} = \frac{x+1}{(y-2)^2}$$

(4) $\sqrt[n]{a^m} = \left(\sqrt[n]{a}\right)^m$

Examples 1.23

$$\sqrt[3]{(27)^4} = (\sqrt[3]{27})^4 = 3^4 = 81$$

(5) $\sqrt[m]{\sqrt[n]{a}} = \sqrt[mn]{a}$

Examples 1.24

$$\sqrt[3]{\sqrt{5}} = \sqrt[6]{5}, \qquad \sqrt[4]{\sqrt[3]{2}} = \sqrt[12]{2}, \qquad \sqrt[5]{\sqrt[3]{x^2}} = \sqrt[15]{x^2}$$

Simplifying Radicals

The form of a radical may be changed in the following ways:

(1)   Removal of perfect nth powers from the radicand.

Examples 1.25

$$\sqrt[3]{32} = \sqrt[3]{2^3(4)} = \sqrt[3]{2^3} \cdot \sqrt[3]{4} = 2\sqrt[3]{4}$$

$$\sqrt{8x^5y^7} = \sqrt{(4x^4y^6)(2xy)} = \sqrt{4x^4y^6}\,\sqrt{2xy} = 2x^2y^3\sqrt{2xy}$$

(2)   Reduction of the index of the radical.

Examples 1.26

$\sqrt[4]{64} = \sqrt[4]{2^6} = 2^{6/4} = 2^{3/2} = \sqrt{2^3} = \sqrt{8}$ where the index is reduced from 4 to 2.

$\sqrt[6]{25x^6} = \sqrt[6]{(5x^3)^2} = (5x^3)^{2/6} = (5x^3)^{1/3} = \sqrt[3]{5x^3} = x\sqrt[3]{5}$,

where the index is reduced from 6 to 3.

Note. $\sqrt[4]{(-4)^2} = \sqrt[4]{16} = 2$.

It is *incorrect* to write $\sqrt[4]{(-4)^2} = (-4)^{2/4} = (-4)^{1/2} = \sqrt{-4}$.

(3)   Rationalization of the denominator in the radicand.

Examples 1.27

$$\sqrt[3]{\frac{9}{2}} = \sqrt[3]{\frac{9}{2}\left(\frac{2^2}{2^2}\right)} = \sqrt[3]{\frac{9(2^2)}{2^3}} = \frac{\sqrt[3]{36}}{2}$$

Important Point!

A radical is said to be in its simplest form if:

(a) all perfect *n*th powers have been removed from the radical,

(b) the index of the radical is as small as possible,

(c) no fractions are present in the radicand, i.e., the denominator has been rationalized.

Complex Numbers

A complex number is an expression of the form a + bi, where a and b are real numbers and $i = \sqrt{-1}$. In the complex number, a is called the real part and bi is the imaginary part.

- Two complex numbers a + bi and c + di are equal if and only if a = c and b = d.

- The complex number a + bi = 0 if and only if a = 0, b = 0.

- The complex number c + di is real if d = 0. If c + di = 3, then c = 3, d = 0.

- The conjugate of a complex number a + bi is a - bi, and conversely. Thus, 5 - 3i and 5 + 3i are conjugates.

Graphical Representation of Complex Numbers

Employing rectangular coordinate axes, the complex number x + yi is represented by, or corresponds to, the point whose coordinates are (x, y). See Figure 1-2.

- To represent the complex number 3 + 4i, measure off 3 units distance along X'X and to the right of O, and then up 4 units distance.

- To represent the number -2 + 3i, measure off 2 units distance along X'X and to the left of O, and then up 3 units distance.

- To represent the complex number -1 - 4i, measure off 1 unit distance along X'X and to the left of O, and then down 4 units distance.

- To represent the number 2 - 4i, measure off 2 units distance along X'X and to the right of O, and then down 4 units distance.

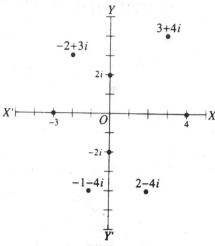

Figure 1-2

Algebraic Operations with Complex Numbers

- To add two complex numbers, add the real parts and the imaginary parts separately.

  (a + bi) + (c + di) = (a + c) + (b + d)i
  (5 + 4i) + (3 + 2i) = (5 + 3) + (4 + 2)i = 8 + 6i
  (- 6 + 2i) + (4 - 5i) = (- 6 + 4) + (2 - 5)i = - 2 - 3i

- To subtract two complex numbers, subtract the real parts and the imaginary parts separately.

  (a + bi) - (c + di) = (a - c) + (b - d)i
  (3 + 2i) - (5 - 3i) = (3 - 5) + (2 + 3)i = - 2 + 5i
  (- 1 + i) - (- 3 + 2i) = (- 1 + 3) + (1 - 2)i = 2 - i

- To multiply two complex numbers, treat the numbers as ordinary binomials and replace $i^2 = -1$.

$$(a + bi)(c + di) = ac + adi + bci + bdi^2 = (ac - bd) + (ad + bc)i$$
$$(5 + 3i)(2 - 2i) = 10 - 10i + 6i - 6i^2 = 10 - 4i - 6(-1) = 16 - 4i$$

- To divide two complex numbers, multiply the numerator and denominator of the fraction by the conjugate of the denominator, replacing $i^2 = -1$.

$$\frac{2+i}{3-4i} = \left(\frac{2+i}{3-4i}\right)\left(\frac{3+4i}{3+4i}\right) = \frac{6+8i+3i+4i^2}{9-16i^2} = \frac{2+11i}{25} = \frac{2}{25} + \frac{11}{25}i$$

Chapter 2
ALGEBRAIC EXPRESSIONS AND OPERATIONS

IN THIS CHAPTER:

✔ *Algebraic Expressions*
✔ *Special Products*
✔ *Products Yielding Answers of the Form $a^n \pm b^n$*
✔ *Factoring*
✔ *Factorization Procedures*
✔ *Greatest Common Factor*
✔ *Least Common Multiple*
✔ *Algebraic Fractions*
✔ *Operations with Algebraic Fractions*
✔ *Complex Fractions*

Algebraic Expressions

An **algebraic expression** is a combination of ordinary numbers and letters which represent numbers.

Examples 2.1 $3x^2 - 5xy + 2y^4$, $2a^3b^5$, and

$$\frac{5xy + 3z}{2a^3 - c^2}$$

are algebraic expressions.

A **term** consists of products and quotients of ordinary numbers and letters which represent numbers.

A **monomial** is an algebraic expression consisting of only one term, and a **multinomial** consists of more than one term. More specifically, a **binomial** is an algebraic expression consisting of two terms and a **trinomial** consists of three terms.

Examples 2.2

$7x^3y^4$, $4x^2/y$	are monomials
$2x+4y$, $3x^4 - 4xyz^3$	are binomials
$3x^2 -5x + 2$, $x^3 - 3xy/z - 2x^3z^7$	are trinomials
$7x + 6y$, $7x + 5x^2/y - 3x^3/16$	are multinomials

Terms

One factor of a term is said to be the coefficient of the rest of the term. Thus, in the term $5x^3y^2$, $5x^3$ is the coefficient of y^2, $5y^2$ is the coefficient of x^3, and 5 is the coefficient of x^3y^2.

Two or more like terms in an algebraic expression may be combined into one term. Thus, $7x^2y - 4x^2y + 2x^2y$ may be combined and written $5x^2y$.

A term is integral and rational in certain literals (letters which represent numbers) if the term consists of:

- positive integer powers of the variables multiplied by a factor not containing any variables; or

- no variables at all.

Examples 2.3 The terms $6x^2y^3$, $-5y^4$, 7, $-4x$, and $\sqrt{3}\, x^3y^6$ are integral and rational in the variables present. However, $3\sqrt{x}$ is not rational in x, and $4/x$ is not integral in x.

A polynomial is a monomial or multinomial in which every term is integral and rational.

Examples 2.4 $3x^2y^3 - 5x^4y + 2$, $4xy + z$, $2x^4 - 7x^3 + 3x^2 - 5x + 2$, and $3x^2$ are polynomials. However, $3x^2 - 4/x$ and $4\sqrt{y} + 3$ are not polynomials.

Degree

The **degree** of a monomial is the sum of all the exponents in the variables in the term. Thus, the degree of $4x^3y^2z$ is $3 + 2 + 1 = 6$. The degree of a constant, such as 6, 0, $\sqrt{3}$, or π, is zero.

The degree of a polynomial is the same as that of the term having highest degree and non-zero coefficient. Thus, $7x^3y^2 - 4xz^5 + 2x^3y$ has terms of degree 5, 6, and 4, respectively; hence, the degree of the polynomial is 6.

You Need to Know ✔

Grouping

A symbol of grouping such as parentheses (), brackets [], or braces {} is often used to show that the terms contained in them are considered as a single quantity.

Computation with Algebraic Expressions

Addition of algebraic expressions is achieved by combining like terms. In order to accomplish this addition, the expressions may be arranged in rows with like terms in the same column; these columns are then added.

Example 2.5 Add $7x + 3y^3 - 4xy$, $3x - 2y^3 + 7xy$, and $2xy - 5x - 6y^3$.

Write:

$7x$	$3y^3$	$-4xy$
$3x$	$-2y^3$	$7xy$
$-5x$	$-6y^3$	$2xy$

Addition: $\quad 5x \qquad\qquad -5y^3 \qquad\qquad 5xy$

Hence the result is: $\quad 5x - 5y^3 + 5xy$

Subtraction of two algebraic expressions is achieved by changing the sign of every term in the expression which is being subtracted (sometimes called the subtrahend) and adding this result to the other expression (called the minuend).

Example 2.6 Subtract $2x^2 - 3xy + 5y^2$ from $10x^2 - 2xy - 3y^2$.

Write:

	$10x^2$	$-2xy$	$-3y^2$
	$-2x^2$	$+3xy$	$-5y^2$

Subtraction: $8x^2$ $+xy$ $-8y^2$

Hence the result is: $8x^2 + xy - 8y^2$.

Multiplication of algebraic expressions is achieved by multiplying the terms in the factors of the expressions.

(1) To multiply two or more monomials: Use the laws of exponents, the rules of signs, and the commutative and associative properties of multiplication.

Example 2.7 Multiply $-3x^2y^3z$, $2x^4y$, and $-4xy^4z^2$.

- Write: $(-3x^2y^3z)(2x^4y)(-4xy^4z^2)$

- Arranging according to the commutative and associative laws:

  $\{(-3)(2)(-4)\}\{(x^2)(x^4)(x)\}\{(y^3)(y)(y^4)\}\{(z)(z^2)\}$

- Combine using rules of signs and laws of exponents to obtain:

  $24x^7y^8z^3$

(2) To multiply a polynomial by a monomial: Multiply each term of the polynomial by the monomial and combine results.

Example 2.8 Multiply $3xy - 4x^3 + 2xy^2$ by $5x^2y^4$.

- Write: $(5x^2y^4)(3xy - 4x^3 + 2xy^2)$

- Multiply each term:

  $(5x^2y^4)(3xy) + (5x^2y^4)(-4x^3) + (5x^2y^4)(2xy^2)$

- The result is:

  $15x^3y^5 - 20x^5y^4 + 10x^3y^6$

(3) To multiply a polynomial by a polynomial: Multiply each of the terms of one polynomial by each of the terms of the other polynomial and combine results. (It is often very useful to arrange the polynomials according to ascending (or descending) powers of one of the letters involved.)

Example 2.9 Multiply $-3x + 9 + x^2$ by $3 - x$.

- Arrange in descending powers of x:

$$x^2 - 3x + 9 \qquad\qquad \textbf{(A)}$$
$$-x + 3$$
$$\overline{}$$

Multiply **(A)** by -x:	$-x^3 + 3x^2 - 9x$
Multiply **(A)** by 3:	$3x^2 - 9x + 27$
Adding:	$-x^3 + 6x^2 - 18x + 27$

(4) To divide a monomial by a monomial: Find the quotient of the numerical coefficients, find the quotients of the variables, and multiply these quotients.

Example 2.10 Divide $24x^4y^2z^3$ by $-3x^3y^4z$.

$$\frac{24x^4y^2z^3}{-3x^3y^4z} = \left(\frac{24}{-3}\right)\left(\frac{x^4}{x^3}\right)\left(\frac{y^2}{y^4}\right)\left(\frac{z^3}{z}\right)$$

$$= (-8)(x)\left(\frac{1}{y^2}\right)(z^2)$$

$$= -\frac{8xz^2}{y^2}$$

(5) To divide a polynomial by a polynomial:

(a) Arrange the terms of both polynomials in descending (or ascending) powers of one of the variables common to both polynomials.

(b) Divide the first term in the dividend by the first term in the divisor. This gives the first term of the quotient.

(c) Multiply the first term of the quotient by the divisor and subtract from the dividend, thus obtaining a new dividend.

(d) Use the dividend obtained in (c) to repeat steps (b) and (c) until a remainder is obtained which is either of degree lower than the degree of the divisor or zero.

(e) The result is written:

$$\frac{\text{dividend}}{\text{divisor}} = \text{quotient} + \frac{\text{remainder}}{\text{divisor}}$$

Example 2.11 Divide $x^2 + 2x^4 - 3x^3 + x - 2$ by $x^2 - 3x + 2$.

Write the polynomials in descending powers of x and arrange the work as follows:

$$\begin{array}{r}
2x^2 + 3x\ + 6 \\
x^2 - 3x + 2\overline{\smash{\big)}2x^4 - 3x^3 +\ x^2 +\ \ x - 2} \\
\underline{2x^4 - 6x^3 + 4x^2} \\
3x^3 - 3x^2 +\ \ x - 2 \\
\underline{3x^3 - 9x^2 + 6x} \\
6x^2 - 5x - 2
\end{array}$$

Special Products

The following are some of the products which occur frequently in mathematics, and the student should become familiar with them as soon as possible. Proofs of these results may be obtained by performing the multiplications.

I.   **Product of a monomial and a binomial**

$a(c + d) = ac + ad$

Example 2.12   Determine the product $3x(2x + 3y)$.

Using **I** with a = 3x, c = 2x, and d = 3y,

$3x(2x + 3y) = (3x)(2x) + (3x)(3y) = 6x^2 + 9xy$

II.   **Product of the sum and the difference of two terms**

$(a + b)(a - b) = a^2 - b^2$

Example 2.13   Determine the product $(2x + 3y)(2x - 3y)$.

Using **II** with a = 2x and b = 3y,

$(2x + 3y)(2x - 3y) = (2x)^2 - (3y)^2 = 4x^2 - 9y^2$

III.   **Square of a binomial**

$$(a + b)^2 = a^2 + 2ab + b^2$$

$$(a - b)^2 = a^2 - 2ab + b^2$$

Examples 2.14   Determine the products (1) $(3x + 5y)^2$ and
(2) $(7x^2 - 2xy)^2$.

(1)   Using **III** with a = 3x and b = 5y,

$$(3x + 5y)^2 = (3x)^2 + 2(3x)(5y) + (5y)^2 = 9x^2 + 30xy + 25y^2$$

(2)   Using **III** with a = $7x^2$ and b = 2xy

$$(7x^2 - 2xy)^2 = (7x^2)^2 - 2(7x^2)(2xy) + (2xy)^2 = 49x^4 - 28x^3y + 4x^2y^2$$

IV.   **Product of two binomials**

$$(x + a)(x + b) = x^2 + (a + b)x + ab$$

$$(ax + b)(cx + d) = acx^2 + (ad + bc)x + bd$$

$$(a + b)(c + d) = ac + bc + ad + bd$$

Examples 2.15   Determine the products (1) $(x + 3)(x + 5)$ and
(2) $(3x + y)(4x - 2y)$.

(1)   Using **IV** with a = 3 and b = 5,

$$(x + 3)(x + 5) = x^2 + (3 + 5)x + (3)(5) = x^2 + 8x + 15$$

(2)   Using **IV** with a = 3x, b = y, c = 4x, and d = -2y,

$$(3x + y)(4x - 2y) = (3x)(4x) + (y)(4x) + (3x)(-2y) + (y)(-2y)$$
$$= 12x^2 - 2xy - 2y^2$$

V.   Cube of a binomial

$$(a + b)^3 = a^3 + 3a^2b + 3ab^2 + b^3$$

$$(a - b)^3 = a^3 - 3a^2b + 3ab^2 - b^3$$

Examples 2.16   Determine the products (1) $(x + 2y)^3$ and (2) $(2y - 5)^3$.

(1)   Using **V** with a = x and b = 2y,

$$(x + 2y)^3 = x^3 + 3(x)^2(2y) + 3(x)(2y)^2 + (2y)^3$$
$$= x^3 + 6x^2y + 12xy^2 + 8y^3$$

(2)   Using **V** with a = 2y and b = 5,

$$(2y - 5)^3 = (2y)^3 - 3(2y)^2(5) + 3(2y)(5)^2 - (5)^3$$
$$= 8y^3 - 60y^2 + 150y - 125$$

VI.   Square of a trinomial

$$(a + b + c)^2 = a^2 + b^2 + c^2 + 2ab + 2ac + 2bc$$

Example 2.17   Determine the product $(2x + 3y + z)^2$.

Using **VI** with a = 2x, b = 3y, and c = z,

$$(2x + 3y + z)^2 = (2x)^2 + (3y)^2 + (z)^2 + 2(2x)(3y) + 2(2x)(z) + 2(3y)(z)$$
$$= 4x^2 + 9y^2 + z^2 + 12xy + 4xz + 6yz$$

Products Yielding Answers of the Form $a^n \pm b^n$

It may be verified by multiplication that:

$(a - b)(a^2 + ab + b^2) = a^3 - b^3$

$(a - b)(a^3 + a^2b + ab^2 + b^3) = a^4 - b^4$

$(a - b)(a^4 + a^3b + a^2b^2 + ab^3 + b^4) = a^5 - b^5$

$(a - b)(a^5 + a^4b + a^3b^2 + a^2b^3 + ab^4 + b^5) = a^6 - b^6$

etc., the rule being clear. These may be summarized by:

VII. $(a - b)(a^{n-1} + a^{n-2}b + a^{n-3}b^2 + ... + ab^{n-2} + b^{n-1}) = a^n - b^n$

where n is any positive integer (1, 2, 3, 4, ...).

Example 2.18   Determine the product $(x - 2y)(x^2 + 2xy + 4y^2)$.

Using **VII** with a = x and b = 2y,

$(x - 2y)(x^2 + 2xy + 4y^2) = x^3 - (2y)^3 = x^3 - 8y^3$

Similarly, it may be verified that:

$(a + b)(a^2 - ab + b^2) = a^3 + b^3$

$(a + b)(a^4 - a^3b + a^2b^2 - ab^3 + b^4) = a^5 + b^5$

$(a + b)(a^6 - a^5b + a^4b^2 - a^3b^3 + a^2b^4 - ab^5 + b^6) = a^7 + b^7$

etc., the rule being clear. These may be summarized by:

VIII. $(a + b)(a^{n-1} - a^{n-2}b + a^{n-3}b^2 - \ldots - ab^{n-2} + b^{n-1}) = a^n + b^n$

where n is any positive odd integer (1, 3, 5, 7, ...).

Example 2.19   Determine the product $(xy + 2)(x^2y^2 - 2xy + 4)$.

Using **VIII** with a = xy and b = 2,

$$(xy + 2)(x^2y^2 - 2xy + 4) = (xy)^3 + (2)^3 = x^3y^3 + 8$$

Factoring

The factors of a given algebraic expression consist of two or more algebraic expressions which, when multiplied together, produce the given expression.

Examples 2.20 Factor each algebraic expression.

(a). $x^2 - 7x + 6 = (x - 1)(x - 6)$
(b). $x^2 + 8x = x(x + 8)$
(c). $6x^2 - 7x - 5 = (3x - 5)(2x + 1)$
(d). $x^2 + 2xy - 8y^2 = (x + 4y)(x - 2y)$

A **polynomial** is said to be factored completely when it is expressed as a product of prime factors.

* In factoring, we shall allow trivial changes in sign. Thus, $x^2 - 7x + 6$ can be factored either as $(x - 1)(x - 6)$ or $(1 - x)(6 - x)$.

* A polynomial is said to be prime if it has no factors other than plus or minus itself and ± 1.

* Occasionally, we may factor polynomials with rational coefficients, e.g., $x^2 - 9/4 = (x + 3/2)(x - 3/2)$.

- At times, we may factor an expression over a specific set of numbers, e.g., $x^2 - 2 = (x + \sqrt{2})(x - \sqrt{2})$ over the set of real numbers, but it is prime over the set of rational numbers. Unless the set of numbers to use for the coefficients of the factors is specified, it is assumed to be the set of integers.

Factorization Procedures

The following procedures in factoring are very useful.

A. Common monomial factor.
   Type: ac + ad = a(c + d)

Examples 2.21   (a) $6x^2y - 2x^3 = 2x^2(3y - x)$
                    (b) $2x^3y - xy^2 + 3x^2y = xy(2x^2 - y + 3x)$

B. Difference of two squares.
   Type: $a^2 \cdot b^2 = (a + b)(a - b)$

Examples 2.22   (a) $x^2 - 25 = x^2 - 5^2 = (x + 5)(x - 5)$
                    where a = x, b = 5
                    (b) $4x^2 - 9y^2 = (2x)^2 - (3y)^2 = (2x + 3y)(2x - 3y)$
                    where a = 2x, b = 3y

C. Perfect square trinomials.
   Types: $a^2 + 2ab + b^2 = (a + b)^2$
   $a^2 - 2ab + b^2 = (a - b)^2$

Examples 2.23   (a) $x^2 + 6x + 9 = (x + 3)^2$
                    (b) $9x^2 - 12xy + 4y^2 = (3x - 2y)^2$

D. Other trinomials.

Types: $x^2 + (a + b)x + ab = (x + a)(x + b)$

$acx^2 + (ad + bc)x + bd = (ax + b)(cx + d)$

Examples 2.24 (a) $x^2 - 5x + 4 = (x - 4)(x - 1)$

where $a = -4$, $b = -1$

(b) $x^2 + xy - 12y^2 = (x - 3y)(x + 4y)$

where $a = -3y$, $b = 4y$

(c) $8 - 14x + 5x^2 = (4 - 5x)(2 - x)$

E. Sum, difference of two cubes.

Types: $a^3 + b^3 = (a + b)(a^2 - ab + b^2)$

$a^3 - b^3 = (a - b)(a^2 + ab + b^2)$

Examples 2.25 (a) $8x^3 + 27y^3 = (2x)^3 + (3y)^3$

$= (2x + 3y)[(2x)^2 - (2x)(3y) + (3y)^2]$

$= (2x + 3y)(4x^2 - 6xy + 9y^2)$

(b) $8x^3y3 - 1 = (2xy)^3 - 1^3$

$= (2xy - 1)(4x^2y^2 + 2xy + 1)$

F. Grouping of terms.

Type: $ac + bc + ad + bd = c(a + b) + d(a + b)$

$= (a + b)(c + d)$

Example 2.26 $2ax - 4bx + ay - 2by = 2x(a - 2b) + y(a - 2b)$

$= (a - 2b)(2x + y)$

G. Factors of $a^n \pm b^n$.

Examples 2.27 (a) $32x^5 + 1 = (2x)^5 + 1^5$

$= (2x + 1)[(2x)^4 - (2x)^3 + (2x)^2 - 2x + 1]$

$= (2x + 1)[16x^4 - 8x^3 + 4x^2 - 2x + 1]$

(b) $x^7 - 1 = (x - 1)(x^6 + x^5 + x^4 + x^3 + x^2 + x + 1)$

H. Addition and subtraction of suitable terms.

Example 2.28   Factor $x^4 + 4$.

Adding and subtracting $4x^2$ (twice the product of the square roots of x^4 and 4), we have

$$x^4 + 4 = (x^4 + 4x^2 + 4) - 4x^2 = (x^2 + 2)^2 - (2x)^2$$
$$= (x^2 + 2)^2 - (2x)^2$$
$$= (x^2 + 2 + 2x)(x^2 + 2 - 2x)$$
$$= (x^2 + 2x + 2)(x^2 - 2x + 2)$$

I. Miscellaneous combinations of previous methods.

Example 2.29 $x^4 - xy^3 - x^3y + y^4 = (x^4 - xy^3) - (x^3y - y^4)$
$$= x(x^3 - y^3) - y(x^3 - y^3)$$
$$= (x^3 - y^3)(x - y) = (x - y)(x^2 + xy + y^2)(x - y)$$
$$= (x - y)^2(x^2 + xy + y^2)$$

Greatest Common Factor

The **greatest common factor** (GCF) of two or more given polynomials is the polynomial of highest degree and largest numerical coefficients (apart from trivial changes in sign) which is a factor of all the given polynomials.

To find the GCF of several polynomials:

• Write each polynomial as a product of prime factors.

• The GCF is the product obtained by taking each factor to the lowest power to which it occurs in any of the polynomials.

Example 2.30   Find the GCF of $2^3 3^2(x - y)^3(x + 2y)^2$, $2^2 3^3(x - y)^2(x + 2y)^3$, and $3^2(x - y)^2(x + 2y)$.

The GCF of all three polynomials is: $3^2(x - y)^2(x + 2y)$.

Least Common Multiple

The **least common multiple** (LCM) of two or more given polynomials is the polynomial of lowest degree and smallest numerical coefficients (apart from trivial changes in sign) for which each of the given polynomials will be a factor.

To find the LCM of several polynomials:

- Write each polynomial as a product of prime factors.

- The LCM is the product obtained by taking each factor to the highest power to which it occurs.

Example 2.31   Find the LCM of $2^3 3^2(x - y)^3(x + 2y)^2$, $2^2 3^3(x - y)^2(x + 2y)^3$, and $3^2(x - y)^2(x + 2y)$.

The LCM of all three polynomials is: $2^3 3^3(x - y)^3(x + 2y)^3$.

Algebraic Fractions

Rational Algebraic Fractions

A rational algebraic fraction is an expression which can be written as the quotient of two polynomials, P/Q. P is called the numerator and Q the denominator of the fraction. Thus,

$$\frac{3x-4}{x^2-6x+8} \quad \text{and} \quad \frac{x^3+2y^2}{x^4-3xy+2y^3}$$

are rational algebraic fractions.

Rules for manipulation of algebraic fractions are the same as for fractions in arithmetic. One such fundamental rule is:

The value of a fraction is unchanged if its numerator and denominator are both multiplied by the same quantity or both divided by the same quantity, provided only that this quantity is not zero. In such cases, we call the fractions equivalent.

For example, if we multiply the numerator and denominator of (x + 2)/(x - 3) by (x - 1) we obtain the equivalent fraction

$$\frac{(x+2)(x-1)}{(x-3)(x-1)} = \frac{x^2+x-2}{x^2-4x+3}$$

provided (x - 1) is not zero, i.e., x ≠ 1.

Similarly, given the fraction $(x^2 + 3x + 2)/(x^2 + 4x + 3)$ we may write it as:

$$\frac{(x+2)(x+1)}{(x+3)(x+1)}$$

and divide numerator and denominator by (x + 1) to obtain (x + 2)/(x + 3) provided (x + 1) is not zero, i.e., x ≠ - 1.

Three signs are associated with a fraction: the sign of the numerator, of the denominator, and of the entire fraction. Any two of these signs may be changed without changing the value of the fraction. If there is no sign before a fraction, a plus sign is implied.

Examples 2.32

$$\frac{-a}{b} = \frac{a}{-b} = -\frac{a}{b}, \quad \frac{-a}{-b} = \frac{a}{b}, \quad -\left(\frac{-a}{-b}\right) = -\frac{a}{b}$$

Change of sign may often be of use in simplification. Thus

$$\frac{x^2-3x+2}{2-x}=\frac{(x-2)(x-1)}{2-x}=\frac{(x-2)(x-1)}{-(x-2)}=\frac{x-1}{-1}=1-x$$

Operations with Algebraic Fractions

The algebraic sum of fractions having a common denominator is a fraction whose numerator is the algebraic sum of the numerators of the given fractions and whose denominator is the common denominator.

Example 2.33

$$\frac{2}{x-3}-\frac{3x+4}{x-3}+\frac{x^2+5}{x-3}=\frac{2-(3x+4)+(x^2+5)}{x-3}=\frac{x^2-3x+3}{x-3}$$

To add or subtract fractions having different denominators, write each of the given fractions as equivalent fractions, all having a common denominator.

Example 2.34

$$\frac{2x+1}{x(x+2)}-\frac{3}{(x+2)(x-1)}=\frac{(2x+1)(x-1)-3x}{x(x+2)(x-1)}$$

$$=\frac{(2x+1)(x-1)}{x(x+2)(x-1)}-\frac{3x}{x(x+2)(x-1)}$$

$$=\frac{2x^2-4x-1}{x(x+2)(x-1)}$$

The product of two or more given fractions produces a fraction whose numerator is the product of the numerators of the given fractions and whose denominator is the product of the denominators of the given fractions.

Example 2.35

$$\frac{x^2-9}{x^2-6x+5}\cdot\frac{x-5}{x+3}=\frac{(x+3)(x-3)}{(x-5)(x-1)}\cdot\frac{x-5}{x+3}=\frac{x-3}{x-1}$$

The quotient of two given fractions is obtained by inverting the divisor and then multiplying.

Example 2.36

$$\frac{7}{x^2-4}\div\frac{xy}{x+2}=\frac{7}{(x+2)(x-2)}\cdot\frac{x+2}{xy}=\frac{7}{xy(x-2)}$$

Complex Fractions

A complex fraction is one which has one or more fractions in the numerator or denominator, or in both. To simplify a complex fraction:

Method I

* Reduce the numerator and denominator to simple fractions.

* Divide the two resulting fractions.

Example 2.37

$$\frac{x-\dfrac{1}{x}}{1+\dfrac{1}{x}}=\frac{\dfrac{x^2-1}{x}}{\dfrac{x+1}{x}}=\frac{x^2-1}{x}\cdot\frac{x}{x+1}=\frac{x^2-1}{x+1}=x-1$$

Method II

- Multiply the numerator and denominator of the complex fraction by the LCM of all denominators of the fractions in the complex fraction.

- Reduce the resulting fraction to lowest terms.

Example 2.38

$$\frac{\dfrac{1}{x^2}-4}{\dfrac{1}{x}-2}=\frac{\left(\dfrac{1}{x^2}-4\right)x^2}{\left(\dfrac{1}{x}-2\right)x^2}=\frac{1-4x^2}{x-2x^2}=\frac{(1+2x)(1-2x)}{x(1-2x)}=\frac{1+2x}{x}$$

Chapter 3
FUNCTIONS

IN THIS CHAPTER:

✔ *Ratio, Proportion, and Variation*
✔ *Functions and Graphs*
✔ *Polynomial Functions*
✔ *Rational Functions*
✔ *Partial Fractions*

Ratio, Proportion, and Variation

Ratio

The **ratio** of two numbers a and b, written a:b, is the fraction a/b, provided b ≠ 0. If a = b ≠ 0, the ratio is 1:1 or 1/1 = 1.

Examples 3.1

(a)   The ratio of 4 to 6 = 4:6 = $\dfrac{4}{6} = \dfrac{2}{3}$

(b) $\dfrac{2}{3} : \dfrac{4}{5} = \dfrac{2/3}{4/5} = \dfrac{5}{6}$

(c) $5x : \dfrac{3y}{4} = \dfrac{5x}{3y/4} = \dfrac{20x}{3y}$

Proportion

A **proportion** is an equality of two ratios. Thus a:b – c:d, or a/b – c/d, is a proportion in which a and d are called the extremes and b and c the means, while d is called the fourth proportional to a, b, and c. In the proportion a:b = b:c, c is called the third proportional to a and b, and b is called a mean proportional between a and c.

Proportions are equations and can be transformed using procedures for equations. Some of the transformed equations are used frequently and are called the laws of proportion. If a/b = c/d, then

$(1)\ ad = bc$ $(2)\ \dfrac{b}{a} = \dfrac{d}{c}$ $(3)\ \dfrac{a}{c} = \dfrac{b}{d}$ $(4)\ \dfrac{a+b}{b} = \dfrac{c+d}{d}$

$(5)\ \dfrac{a-b}{b} = \dfrac{c-d}{d}$ $(6)\ \dfrac{a+b}{a-b} = \dfrac{c+d}{c-d}$

Examples 3.2 Find the ratio of each of the following quantities.

(a)   6 pounds to 12 ounces
       It is customary to express the quantities in the same units. Then
       the ratio of 96 ounces to 12 ounces is 96:12 = 8:1.

(b)   3 quarts to 2 gallons
       The required ratio is 3 quarts to 8 quarts or 3:8.

(c)   3 square yards to 6 square feet.
       Since 1 square yard = 9 square feet, the required ratio is 27
       ft^2:6 ft^2 = 9:2.

Example 3.3 A line segment 30 inches long is divided into two parts
whose lengths have the ratio 2:3. Find the lengths of the parts.

Let the required lengths be x and 30 - x. Then,

$$\frac{x}{30-x} = \frac{2}{3}$$

Solving for x,

x = 12 in and 30 - x = 18 in.

Variation

In reading scientific material, it is common to find such statements as
"The pressure of an enclosed gas varies directly as the temperature."
This and similar statements have precise mathematical meanings and
they represent a specific type of function called variation functions. The
three general types of variation functions are direct, inverse, and joint.

(1)   If x varies **directly** as y, then x = ky or x/y = k, where k is
       called the constant of proportionality or the constant of variation.

(2)   If x varies **directly** as y^2, then $x = ky^2$.

(3)   If x varies **inversely** as y, then $x = k/y$.

(4)   If x varies **inversely** as y^2, then $x = k/y^2$.

(5)   If x varies **jointly** as y and z, then $x = kyz$.

(6)   If x varies **directly** as y^2 and **inversely** as z, then $x = ky^2/z$.

The constant k may be determined if one set of values of the variables is known.

Example 3.4 The kinetic energy E of a body is proportional to its weight W and to the square of its velocity v. An 8 lb body moving at 4 ft/sec has 2 ft-lb of kinetic energy. Find the kinetic energy of a 3 ton (6000 lb) truck speeding at 60 mi/hr (88 ft/sec).

To find k: $E = kWv^2$ or

$$k = \frac{E}{Wv^2} = \frac{2 \text{ ft} - \text{lb}}{(8 \text{ lb})(4 \text{ ft/sec})^2} = \frac{1}{64} \text{sec}^2$$

Thus, the kinetic energy of the truck is:

$$E = \frac{Wv^2}{64 \text{ sec}^2} = \frac{(6000 \text{ lb})(88 \text{ ft/sec})^2}{64 \text{ sec}^{-2}} = 726,000 \text{ ft} - \text{lb}$$

Functions and Graphs

Variables

A **variable** is a symbol which may assume any one of a set of values during a discussion. A **constant** is a symbol which stands for only one particular value during the discussion.

Relations

A **relation** is a set of ordered pairs. An ordered pair consists of 2 components or coordinates that define the position of a point in reference to an origin. The relation may be specified by an equation, a rule, or a table. The set of the first components of the ordered pairs is called the **domain** of the relation. The set of the second components is called the **range** of the relation.

Example 3.5 What is the domain and range of the relation {(1, 3), (2, 6), (3, 9), (4, 12)}?

Domain = {1, 2, 3, 4}   Range = {3, 6, 9, 12}

Functions

A **function** is a relation such that each element in the domain is paired with exactly one element in the range.

Examples 3.6 Which relations are functions?

(a)   {(1, 2), (2, 3), (3, 4), (4, 5)}
         Function because each first element is paired with exactly one second element.

(b)   {(1, 2), (1, 3), (2, 8), (3, 9)}
         Not a function because 1 is paired with 2 and with 3.

(c)   {(1, 3), (2, 3), (4, 3), (9, 3)}
         Function because each first element is paired with exactly one second element.

Often, functions and relations are stated as **equations**. When the domain is not stated, we determine the largest subset of the real numbers for which the equation is defined, and that is the domain. Once the domain has been determined, we determine the range by finding the value of the equation for each value of the domain.

Important Point!

The variable associated with the domain is called an **independent variable** and the variable associated with the range is called the **dependent variable**. In equations with the variables x and y, we generally assume that x is the independent variable and that y is the dependent variable.

Example 3.7 What is the domain and range of $y = x^2 + 2$?

The domain is the set of all real numbers since the square of each real number is a real number and that a real number plus 2 is still a real number. Domain = {all real numbers}.

The range is the set of all real numbers greater than or equal to 2 since the square of a real number is at least zero. Also, any real number ≥ 2 is represented in the form $x^2 + 2$. Thus, when 2 is added to each value, we have all the real numbers that are at least 2. Range = {all real numbers ≥ 2}.

Example 3.8 What is the domain and range of y = 1/(x - 3)?

The equation is not defined when x = 3, so the domain is the set of all real numbers not equal to 3. Domain = {real numbers ≠ 3}.

A fraction can be zero only when the numerator can be zero. Since the numerator of this fraction is always 1, the fraction can never equal zero. Thus, the range is the set of all real numbers not equal to 0. Range = {real numbers ≠ 0}.

Function Notation

The notation y = f(x), read "y equals f of x," is used to designate that y is a function of x. With this notation, f(a) represents the value of the dependent variable y when x = a (provided that there is a value).

Thus, y = x^2 - 5x + 2 may be written f(x) = x^2 - 5x + 2. Then, f(2), i.e., the value of f(x) or y when x = 2, is f(2) = 2^2 - 5(2) + 2 = - 4. Similarly, f(-1) = $(-1)^2$ - 5(-1) + 2 = 8.

Any letter may be used in the function notation; thus g(x), h(x), F(x), etc., may represent functions of x.

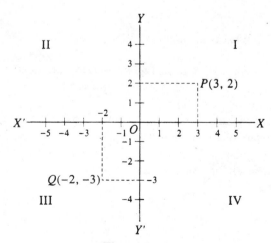

Figure 3-1

Rectangular Coordinate System

A **rectangular coordinate system** is used to give a picture of the relationship between two variables.

Consider two mutually perpendicular lines X'X and Y'Y intersecting in the point O, as shown in Figure 3-1.

The line X'X, called the x axis, is usually horizontal.
The line Y'Y, called the y axis, is usually vertical.
The point O is called the origin.

Using a convenient unit of length, lay off points on the x axis at successive units to the right and left of the origin O, labeling those points to the right 1, 2, 3, 4,... and those to the left -1, -2, -3, -4,....

Here we have arbitrarily chosen OX to be the positive direction; this is customary but not necessary.

Do the same on the y axis, choosing OY as the positive direction. It is customary (but not necessary) to use the same unit of length on both axes.

The x and y axes divide the plane into 4 parts known as **quadrants**, which are labeled I, II, III, IV as in Figure 3-1.

Given a point P in this xy plane, drop perpendiculars from P to the x and y axes. The values of x and y at the points where these perpendiculars meet the x and y axes determine respectively the **x coordinate (abscissa)** of the point and the **y coordinate (or ordinate)** of the point P. These coordinates are indicated by the symbol (x, y).

Conversely, given the coordinates of a point, we may locate or plot the point in the xy plane. For example, the point P in Figure 3-1 has coordinates (3, 2); the point having coordinates (-2, -3) is Q.

The graph of a function y = f(x) is the set of all points (x, y) satisfied by the equation y = f(x).

Symmetry

When the left half of a graph is a mirror image of the right half, we say the graph is symmetric with respect to the y axis (see Figure 3-2).

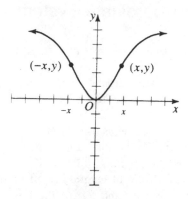

Figure 3-2

This symmetry occurs because for any x value, both x and -x result in the same y value, that is $f(x) = f(-x)$. The equation may or may not be a function for $f(x)$ in terms of x.

Some graphs have a bottom half that is the mirror image of the top half, and we say these graphs are symmetric with respect to the x axis. Symmetry with respect to the x axis results when for each y, both y and -y result in the same x value (see Figure 3-3). In these cases, you do not have a function for y in terms of x.

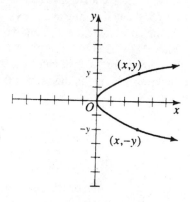

Figure 3-3

If substituting for -x for x and -y for y in an equation yields an equivalent equation, we say the graph is symmetric with respect to the origin (see Figure 3-4). These equations represent relations that are not always functions.

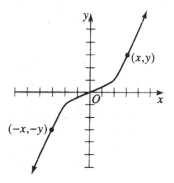

Figure 3-4

Symmetry can be used to make sketching the graphs of relations and functions easier. Once the type of symmetry, if any, and the shape of half of the graph have been determined, the other half of the graph can be drawn by using this symmetry. Most graphs are not symmetric with respect to the y axis, x axis, or the origin. However, many frequently used graphs do display one of these symmetries and using that symmetry in graphing the relation simplifies the graphing process.

Example 3.9 Test the relation y = 1/x for symmetry.

Substituting -x for x, we get y = -1/x, so the graph is not symmetric with respect to the y axis.

Substituting -y for y, we get -y = 1/x, so the graph is not symmetric with respect to the x axis.

Substituting -x for x and -y for y, we get -y = -1/x, which is equivalent to y = 1/x, so the graph is symmetric with respect to the origin.

Shifts

The graph of $y = f(x)$ is shifted upward by adding a positive constant to each y value in the graph. It is shifted downward by adding a negative constant to each y value in the graph of $y = f(x)$. Thus, the graph of $y = f(x) + b$ differs from the graph of $y = f(x)$ by a vertical shift of $|b|$ units. The shift is up if $b > 0$ and the shift is down if $b < 0$.

Example 3.10 How do the graphs of $y = x^2 + 2$ and $y = x^2 - 3$ differ from the graph of $y = x^2$?

The graph of $y = x^2$ is shifted up 2 units to yield the graph of
$y = x^2 + 2$ (see Figures 3-5(a) and (b)).
The graph of $y = x^2$ is shifted 3 units down to yield the graph of
$y = x^2 - 3$ (see Figures 3-5(a) and (c)).

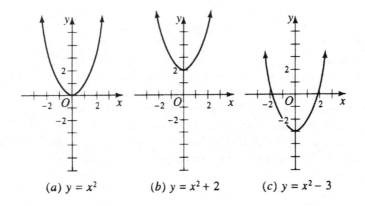

(a) $y = x^2$ (b) $y = x^2 + 2$ (c) $y = x^2 - 3$

Figure 3-5

The graph of $y = f(x)$ is shifted to the right when a positive number is subtracted from each x value. It is shifted to the left if a negative number is subtracted from each x value. Thus, the graph of $y = f(x - a)$ differs from the graph of $y = f(x)$ by a horizontal shift of $|a|$ units. The shift is to the right if $a > 0$ and the shift is to the left if $a < 0$.

Example 3.11 How do the graphs of $y = (x + 1)^2$ and $y = (x - 2)^2$ differ from the graph of $y = x^2$?

The graph of $y = x^2$ is shifted 1 unit to the left to yield the graph of $y = (x + 1)^2$ since $x + 1 = x - (- 1)$ [see Figures 3-6 (a) and (b)]. The graph of $y = x^2$ is shifted 2 units to the right to yield the graph of $y = (x - 2)^2$ (see Figures 3-6 (a) and (c)).

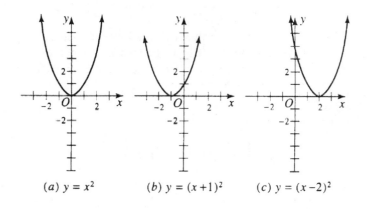

$(a)\ y = x^2$ $(b)\ y = (x+1)^2$ $(c)\ y = (x-2)^2$

Figure 3-6

Scaling

If each y value is multiplied by a positive number greater than 1, the rate of change in y is increased from the rate of change in y values for $y = f(x)$. However, if each y value is multiplied by a positive number between 0 and 1, the rate of change in y values is decreased from the rate of change in y values for $y = f(x)$. Thus, the graph of $y = cf(x)$, where c is a positive number, differs from the graph of $y = f(x)$ by the rate of increase in y. If $c > 1$, the rate of change in y is increased and if $0 < c < 1$, the rate of change in y is decreased.

The graph of $y = f(x)$ is reflected across the x axis when each y value is multiplied by a negative number. So the graph of $y = cf(x)$, where $c < 0$, is the reflection of $y = |c|f(x)$ across the x axis.

Example 3.12 How do the graphs of $y = -|x|$, $y = 3|x|$, and $y = 1/2|x|$ differ from the graph $y = |x|$?

The graph of $y = |x|$ is reflected across the x axis to yield $y = -|x|$ (see Figures 3-7(a) and (b)).

The graph of $y = |x|$ has the y value multiplied by 3 for each x value to yield the graph of $y = 3|x|$ (see Figures 3-7(a) and (c)).

The graph of $y = |x|$ has the y value multiplied by 1/2 for each x value to yield the graph of $y = 1/2|x|$ (see Figures 3-7(a) and (d)).

Polynomial Functions

Polynomial Equations

A rational integral equation of degree n in the variable x is an equation which can be written in the form:

$$a_n x^n + a_{n-1} x^{n-1} + a_{n-2} x^{n-2} + \cdots + a_1 x + a_0 = 0, \ a_n \neq 0$$

where n is a positive integer and $a_0, a_1, a_2, \ldots, a_{n-1}, a_n$ are constants. The coefficient of the highest degree term is called the lead coefficient and a_0 is called the constant term.

Thus, $4x^3 - 2x^2 + 3x - 5 = 0$, $x^2 - \sqrt{2} \ x + 1/4 = 0$, and $x^4 + \sqrt{-3} \ x - 8 = 0$ are rational integral equations in x of degree 3, 2, and 4 respectively. Note that in each equation the exponents of x are positive and integral, and the coefficients are constants (real or complex numbers).

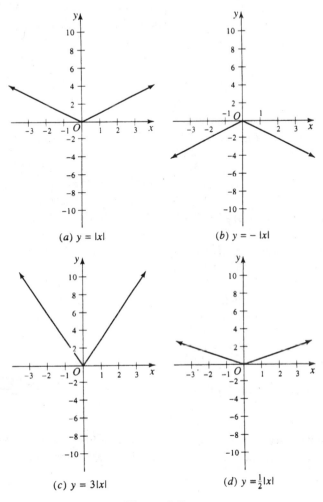

(a) $y = |x|$

(b) $y = -|x|$

(c) $y = 3|x|$

(d) $y = \frac{1}{2}|x|$

Figure 3-7

A **polynomial** of degree n in the variable x is a function of x which can be written in the form

$$P(x) = a_n x^n + a_{n-1} x^{n-1} + a_{n-2} x^{n-2} + \bullet\bullet\bullet + a_1 x + a_0 = 0, \quad a_n \neq 0$$

where n is a positive integer and a_o, a_1, a_2, ..., a_{n-1}, a_n are constants. Then, $P(x) = 0$ is a rational integral equation of degree n in x.

If $P(x) = 3x^3 + x^2 + 5x - 6$,
$$\text{then } P(-2) = 3(-2)^3 + (-2)^2 + 5(-2) - 6 = -36.$$
If $P(x) = x^2 + 2x - 8$,
$$\text{then } P(\sqrt{5}) = 5 + 2\sqrt{5} - 3.$$

Any value of x which makes $P(x)$ vanish is called a **root** of the equation $P(x) = 0$. Thus 2 is a root of the equation $P(x) = 3x^3 + x^2 + 5x - 6 = 0$ since $P(2) = 24 - 8 - 10 - 6 = 0$.

Zeroes of Polynomial Equations

Remainder theorem. If r is any constant and if a polynomial $P(x)$ is divided by $(x - r)$, the remainder is $P(r)$.

For example, if $P(x) = 2x^3 - 3x^2 - x + 8$ is divided by $x + 1$, then $r = -1$ and the remainder $= P(-1) = -2 - 3 + 1 + 8 = 4$. That is,

$$\frac{2x^3 - 3x^2 - x + 8}{x + 1} = Q(x) + \frac{4}{x + 1}$$

where $Q(x)$ is a polynomial in x.

Factor theorem. If r is a root of the equation $P(x) = 0$, i.e., if $P(r) = 0$, then $(x - r)$ is a factor of $P(x)$. Conversely, if $(x - r)$ is a factor of $P(x)$, then r is a root of $P(x) = 0$, or $P(r) = 0$. Thus, 1, -2, -3 are the three roots of the equation $P(x) = x^3 + 4x^2 + x - 6 = 0$, since $P(1) = P(-2) = P(-3) = 0$. Then $(x - 1)$, $(x + 2)$, and $(x + 3)$ are factors of $x^3 + 4x^2 + x - 6$.

Synthetic division. Synthetic division is a simplified method of dividing a polynomial $P(x)$ by $x - r$, where r is any assigned number. By this method, we determine values of the coefficients of the quotient and the value of the remainder can readily be determined.

Example 3.13 Divide $(5x + x^4 - 14x^2)$ by $(x + 4)$ using synthetic division.

Write the terms of the dividend in descending powers of the variable and fill in missing terms using zero for the coefficients; write the divisor in the form x - a.

$$(x^4 + 0x^3 - 14x^2 + 5x + 0) \div (x - (- 4))$$

Write the constant term a from the divisor on the left in a _| and write the coefficients from the dividend to the right of the symbol

-4| 1 + 0 - 14 + 5 + 0

Bring down the first term in the divisor to the third row, leaving a blank row for now.

-4| 1 + 0 - 14 + 5 + 0

     ‾‾‾‾‾‾‾‾‾‾‾‾‾‾
     1

Multiply the term in the quotient row (third row) by the divisor and write the product in the second row under the second term in the first row, add the numbers in the column formed, and write the sum as the second term in the quotient row.

-4| 1 + 0 - 14 + 5 + 0

       - 4
     ‾‾‾‾‾‾‾‾‾‾‾‾‾
     1 - 4

Multiply the last term on the right in the quotient row by the divisor, write it under the next term in the top row, add, and write the sum in the quotient row. Continue this process until all of the terms in the top row have a number under them.

$-\underline{4}|$ $1 + 0 - 14 + 5 + 0$

$$\underline{-4 + 16 - 8 + 12}$$
$$1 - 4 \ + 2 - 3 + 12$$

The third row is the quotient row with the last term being the remainder. The degree of the quotient polynomial is one less than the degree of the dividend because we are dividing by a linear factor. The terms of the quotient row are the coefficients of the terms in the quotient polynomial. The degree of the quotient polynomial here is 3.

The quotient with remainder for $(x^4 + 0x^3 - 14x^2 + 5x + 0) \div (x - (- 4))$ is

$$1x^3 - 4x^2 + 2x - 3 + \frac{12}{x+4}$$

Fundamental theorem of algebra. Every polynomial equation $P(x) = 0$ has at least one root, real or complex.

Thus, $x^7 - 3x^5 + 2 = 0$ has at least one root.

But $f(x) = \sqrt{x} + 3 = 0$ has no root, since no number r exists such that $f(r) = 0$. Since this equation is not rational, the fundamental theorem does not apply.

Number of roots of an equation. Every rational integral equation $P(x) = 0$ of the nth degree has exactly n roots.

Thus, $2x^3 + 5x^2 - 14x - 8 = 0$ has exactly 3 roots, namely 2, -1/2. -4.

Some of the n roots may be equal.

Thus the equation of the sixth degree $(x - 2)^3(x - 5)^2(x + 4) = 0$ has 2 as a triple root, 5 as a double root, and -4 as a single root; i.e., the six roots are 2, 2, 2, 5, 5, -4.

Solving Polynomial Equations

Complex and irrational roots.

(1) If a complex number $a + bi$ is a root of the rational integral equation $P(x) = 0$ with **real coefficients**, then the conjugate complex number $a - bi$ is also a root. It follows that every rational integral equation of odd degree with real coefficients has at least one real root.

(2) If the rational integral equation $P(x) = 0$ with **rational coefficients** has $a + \sqrt{b}$ as a root, where a and b are rational and $\sqrt{b}$ is irrational, then $a - \sqrt{b}$ is also a root.

Rational root theorem. If b/c, a rational fraction in lowest terms, is a root of the equation

$$a_n x^n + a_{n-1} x^{n-1} + a_{n-2} x^{n-2} + \cdots + a_1 x + a_0 = 0, \ a_n \neq 0$$

with integral coefficients, then b is a factor of a_0 and c is a factor of a_n. Thus, if b/c is a rational root of $6x^3 + 5x^2 - 3x - 2 = 0$, the values of b are limited to the factors of 2, which are $\pm1, \pm2$; and the values of c are limited to the factors of 6, which are $\pm1, \pm2, \pm3, \pm6$. Hence the only possible rational roots are $\pm1, \pm2, \pm1/2, \pm1/3, \pm1/6, \pm2/3$.

Integral root theorem. It follows that if an equation $P(x) = 0$ has integral coefficients and the lead coefficient is 1:

$$x^n + a_{n-1}x^{n-1} + a_{n-2}x^{n-2} + \cdots + a_1x + a_0 = 0,$$

then any rational root of $P(x) = 0$ is an integer and a factor of a_0. Thus, the rational roots, if any, of $x^3 + 2x^2 - 11x - 12 = 0$ are limited to the integral factors of 12, which are $\pm1, \pm2, \pm3, \pm4, \pm6, \pm12$.

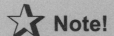 **Note!**

Intermediate value theorem

If $P(x) = 0$ is a polynomial equation with real coefficients, then approximate values of the real roots of $P(x) = 0$ may be found by obtaining the graph of $y = P(x)$ and determining the values of x at the points where the graph intersects the x-axis ($y = 0$). Fundamental in this procedure is the fact that if $P(a)$ and $P(b)$ have opposite signs then $P(x) = 0$ has at least one root between $x = a$ and $x = b$. This fact is based on the continuity of the graph of $y = P(x)$ when $P(x)$ is a polynomial with real coefficients.

Example 3.14 For each real zero of $P(x) = 2x^3 - 5x^2 - 6x + 4$, isolate the zero between two consecutive integers.

Since $P(x) = 2x^3 - 5x^2 - 6x + 4$ has degree 3, there are at most 3 real zeros. We will look for the real zeros in the interval -5 to 5. The interval is arbitrary and may need to be expanded if the real zeros are not found here. By synthetic division, we will find the value of $P(x)$ for each integer in the interval selected. The remainders from the synthetic division are the values of $P(x)$ and are summarized in the table below.

x	- 5	- 4	- 3	- 2	-1	0	1	2	3	4	5
P(x)	-341	-180	-77	-20	3	4	-5	-12	-5	28	99

Note that P(-2) = -20 and P(-1) = 3 have opposite signs, so from the Intermediate Value Theorem there is a real zero between -2 and -1. Similarly, since P(0) = 4 and P(1) = -5, there is a real zero between 0 and 1, and since P(3) = -5 and P(4) = 28, there is a real zero between 3 and 4. Three real zeros have been isolated, so we have located all the real zeros of P(x).

Upper and lower limits for the real roots. A number a is called an **upper limit** or **upper bound** for the real roots of P(x) = 0 if no root is greater than a. A number b is called a **lower limit** or **lower bound** for the real roots of P(x) = 0 if no root is less than b. The following theorem is useful in determining upper and lower limits.

Let $a_nx^n + a_{n-1}x^{n-1} + a_{n-2}x^{n-2} + \bullet\bullet\bullet + a_1x + a_0 = 0$, where $a_0, a_1, \ldots,$ an are real and $a_n > 0$. Then:

* If upon synthetic division of P(x) by x - a, where a ≥ 0, all of the numbers obtained in the third row are positive or zero, then a is an upper limit for all the real roots of P(x) = 0.

* If upon synthetic division of P(x) by x - b, where b ≤ 0, all of the numbers obtained in the third row are alternately positive and negative (or zero), then b is a lower limit for all the real roots of P(x) = 0.

Example 3.15 Find an interval that contains all the real zeros of P(x) = $2x^3 - 5x^2 + 6$.

We will find the integer, b, that is the least upper limit of the real zeros of P(x) and the integer, a, that is the great lower limit on the real zeros of P(x). All real zeros will be in the interval [a, b]. To find a and b, we use the synthetic division on P(x) = $2x^3 - 5x^2 + 6$.

<u>1</u>| 2 - 5 + 0 + 6   <u>2</u>| 2 - 5 + 0 + 6   <u>3</u>| 2 - 5 + 0 + 6

    + 2 - 3 - 3   + 4 - 2 - 4   + 6 + 3 + 9

    2 - 3 - 3 + 3   2 - 2 - 2 + 2   2 + 1 + 3 + 15

When we divide using 3, the quotient row is all positive, so 3 is the smallest integer that is an upper limit for the real zeros of $P(x)$. Thus b = 3.

<u>-1</u>| 2 - 5 + 0 + 6

    - 2 + 7 - 7

    2 - 7 + 7 - 1

When we divide using - 1, the quotient row alternates in sign, so -1 is the greatest integer that is a lower limit for the real zeros of $P(x)$. Thus, a = -1.

The real zeros of $P(x) = 2x^3 - 5x^2 + 6$ are in the interval $(-1, 3)$ or $-1 < x < 3$. Since $P(-1) \neq 0$ and $P(3) \neq 0$, we used interval notation that indicates that neither endpoint is a zero.

Descartes' Rule of Signs. If the terms of a polynomial $P(x)$ with real coefficients are arranged in order of descending powers of x, a variation of sign occurs when two consecutive terms differ in sign. For example, $x^3 - 2x^2 + 3x - 12$ has 3 variations of sign, and $2x^7 - 6x^5 - 4x^4 + x^2 - 2x + 4$ has 4 variations of sign.

Descartes' Rule of Signs says that the number of positive roots of $P(x) = 0$ is either equal to the number of variations of sign of $P(x)$ or is less than that number by an even integer. The number of negative roots of $P(x) = 0$ is either equal to the number of variations of sign of $P(-x)$ or is less than that number by an even integer.

Thus, in $P(x) = x^9 - 2x^5 + 2x^2 - 3x + 12 = 0$, there are 4 variations of sign of $P(x)$; hence the number of positive roots of $P(x) = 0$ is 4, (4 - 2) or (4 - 4). Since $P(-x) = (-x)^9 - 2(-x)^5 + 2(-x)^2 - 3(-x) + 12 = -x^9 + 2x^5$

$+ 2x^2 + 3x + 12 = 0$ has one variation of sign, then $P(x) = 0$ has exactly one negative root. Hence, there are 4, 2, or 0 positive roots, 1 negative root, and at least $9 - (4 + 1) = 4$ complex roots.

Approximating Real Zeros

In solving a polynomial equation $P(x) = 0$, it is not always possible to find all the zeros by the previous methods. We have been able to determine the irrational and imaginary zeros when we were able to find quadratic factors that we could solve using the quadratic equation formula (*See* Chap. 5). If we cannot find the quadratic factors of $P(x) = 0$, we will not be able to solve for the imaginary zeros, but we can often find an approximation for some of the real zeros.

To approximate a real zero of $P(x) = 0$, we must first find an interval that contains a real zero of $P(x) = 0$. We can do this using the Intermediate Value Theorem to locate to numbers a and b such that $P(a)$ and $P(b)$ have opposite signs. We keep using the Intermediate Value Theorem until we have isolated the real zero in an interval small enough that it will be known to the desired degree of accuracy.

Example 3.16 Find a real zero of $x^3 + 3x + 8 = 0$ correct to two decimal places.

By Descartes' Rule of Signs, $P(x) = x^3 + 3x + 8$ has no positive real zeros and 1 negative real zeros.

Using synthetic division, we find $P(-2) = -6$ and $P(-1) = 4$, so by Intermediate Value Theorem, $P(x) = x^3 + 3x + 8$ has a real zero between -2 and -1. We now use synthetic division and the Intermediate Value Theorem to determine the tenths interval containing the zero. The results are summarized in the table below.

x	- 1.0	- 1.1	- 1.2	- 1.3	- 1.4	- 1.5
P(x)	4	3.37	2.67	1.90	1.06	1.13

- 1.6	- 1.7	- 1.8	- 1.9	- 2.0
- 0.80	- 2.01	- 3.23	- 4.56	- 6

We can see that P(- 1.5) is positive and P(- 1.6) is negative so the zero is between - 1.6 and - 1.5. Now, we check for the hundredths digit by using synthetic division on the interval between - 1.6 and - 1.5. We do not have to find all the hundredths values, just a sign change between two consecutive values.

x	- 1.50	- 1.51	- 1.52
P(x)	0.13	0.03	- 0.07

We see that P(- 1.51) is positive and P(- 1.52) is negative, so by the Intermediate Value Theorem, there is a real zero between - 1.51 and - 1.52. Since the real zero is located between - 1.51 and - 1.52, we just need to determine whether it rounds off to - 1.51 or - 1.52. To do this, we find P(- 1.515), which is about - 0.02. This value of P(- 1.515) is negative and P(- 1.51) is positive, so we know that the zero is between - 1.515 and - 1.510, and all the numbers in this interval rounded to two decimal places are - 1.51. Thus, to two decimal places, the only real zero of $x^3 + 3x + 8 = 0$ is - 1.51.

Rational Functions

Rational Functions

A rational function is the ratio of two polynomial functions. If P(x) and Q(x) are polynomials, then a function of the form R(x) = P(x)/Q(x) is a rational function where Q(x) ≠ 0. The domain of R(x) is the intersection of the domains of P(x) and Q(x).

Vertical Asymptotes

If R(x) = P(x)/Q(x) , then values of x that make Q(x) = 0 result in vertical asymptotes if P(x) ≠ 0. However, if for some value x = a, P(a) = 0 and Q(a) = 0, then P(x) and Q(x) have a common factor of x - a. If R(x) is then reduced to lowest terms, the graph of R(x) has a hole in it where x = a.

A vertical asymptote for R(x) is a vertical line x = k, k being a constant, that the graph of R(x) approaches but does not touch. R(k) is not defined because Q(k) = 0 and P(k) ≠ 0. The domain of R(x) is separated into distinct intervals by the vertical asymptotes of R(x).

Example 3.17 What are the vertical asymptotes of

$$R\left(x\right)=\frac{2x-3}{x^2-4}$$

Since R(x) is undefined when x^2 - 4 = 0, x = 2 and x = -2 could result in vertical asymptotes. When x = 2, 2x - 3 ≠ 0 and when x = -2, 2x - 3 ≠ 0. Thus, the graph of R(x) has vertical asymptotes of x = 2 and x = -2.

Horizontal Asymptotes

A rational function $R(x) = P(x)/Q(x)$ has a horizontal asymptote $y = a$ if, as $|x|$ increases without limit, $R(x)$ approaches a. $R(x)$ has at most one horizontal asymptote. The horizontal asymptote of $R(x)$ may be found from a comparison of the degree of $P(x)$ and the degree of $Q(x)$.

- If the degree of $P(x)$ is less than the degree of $Q(x)$, then $R(x)$ has a horizontal asymptote of $y = 0$.

- If the degree of $P(x)$ is equal to the degree of $Q(x)$, then $R(x)$ has a horizontal asymptote of $y = a_n/b_n$, where a_n is the lead coefficient (coefficient of the highest degree term) of $P(x)$ and b_n is the lead coefficient of $Q(x)$.

- If the degree of $P(x)$ is greater than the degree of $Q(x)$, then $R(x)$ does not have a horizontal asymptote.

The graph of $R(x)$ may cross a horizontal asymptote in the interior of its domain. This is possible since we are only concerned with how $R(x)$ behaves as $|x|$ increases without limit in determining the horizontal asymptote.

Example 3.18 What are the horizontal asymptotes of each rational function $R(x)$?

$$(a)\ R(x) = \frac{3x^3}{x^2 - 1};\ (b)\ R(x) = \frac{x}{x^2 - 4};\ (c)\ R(x) = \frac{2x + 1}{3 + 5x}$$

(a) The degree of the numerator $3x^3$ is 3 and the degree of the denominator is 2. Since the numerator exceeds the degree of the denominator, $R(x)$ does not have a horizontal asymptote.

(b) The degree of the numerator is 1 and the degree of the denominator is 2, so $R(x)$ has a horizontal asymptote of $y = 0$.

(c) The numerator and denominator each have degree 1. Since the lead coefficient of the numerator is 2 and the lead coefficient of the denominator is 5, R(x) has a horizontal asymptote of y = 2/5.

Graphing Rational Functions

To graph a rational function R(x) = P(x)/Q(x), we first determine the holes: values of x for which both P(x) and Q(x) are zero. After any holes are located, we reduce R(x) to lowest terms. The value of the reduced form of R(x) for an x that yields a hole is the y coordinate of the point that corresponds to the hole.

   Once R(x) is in lowest terms, we determine the asymptotes, symmetry, zeros, and y intercept if they exist. We graph the asymptotes as dashed lines, plot the zeros and y intercept, and plot several other points to determine how the graph approaches the asymptotes. Finally, we sketch the graph through the plotted points and approaching the asymptotes.

Example 3.19 Sketch a graph of the rational function,

$$R\left(x\right) = \frac{3}{x^2 - 1}$$

R(x) has vertical asymptotes at x = 1, and at x = -1, a horizontal asymptote of y = 0, and no holes. Since the numerator of R(x) is a constant, it does not have any zeros. Since R(0) = -3, R(x) has a y intercept of (0, -3).

   Plot the y intercept and graph the asymptotes as dashed lines. We determine some values of R(x) in each interval of the domain (-∞, -1), and (-1, 1), and (1, ∞). R(-x) = R(x), so R(x) is symmetric with respect to the y axis.

$$R\left(2\right) = R\left(-2\right) = \frac{3}{2^2 - 1} = 1$$

$$R\left(0.5\right) = R\left(-0.5\right) = \frac{3}{\left(0.5\right)^2 - 1} = -4$$

Plot (2, 1), (-2, 1), (0.5, -4), and (-0.5, -4). Using the asymptotes as a boundary, we sketch the graph, shown in Figure 3-8.

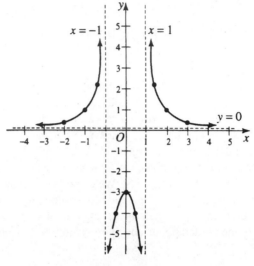

Figure 3-8

Partial Fractions

Rational Fractions

A **rational fraction** in x is the quotient $\dfrac{P(x)}{Q(x)}$ of two polynomials in x. Thus,

$$\frac{3x^2-1}{x^3+7x^2-4}$$

is a rational fraction.

Proper Fractions

A **proper fraction** is one in which the degree of the numerator is less than the degree of the denominator. Thus,

$$\frac{2x-3}{x^2+5x+4} \text{ and } \frac{4x^2+1}{x^4-3x} \text{ are proper fractions.}$$

An **improper fraction** is one in which the degree of the numerator is greater than or equal to the degree of the denominator. Thus,

$$\frac{2x^3+6x^2-9}{x^2-3x+2} \text{ is an improper fraction.}$$

By division, an improper fraction may always be written as the sum of a polynomial and a proper fraction. Thus,

$$\frac{2x^3+6x^2-9}{x^2-3x+2}=2x+12+\frac{32x-33}{x^2-3x+2}$$

Partial Fractions

A given proper fraction may often be written as the sum of other fractions (called partial fractions) whose denominators are of lower degree than the denominator of the given fraction.

Example 3.20

$$\frac{3x-5}{x^2-3x+2}=\frac{3x-5}{(x-1)(x-2)}=\frac{2}{x-1}+\frac{1}{x-2}$$

Fundamental Theorems

A proper fraction may be written as the sum of partial fractions according to the following rules.

(1)   Linear factors none of which are repeated.

If a linear factor ax + b occurs once as a factor of the denominator of the given fraction, then corresponding to this factor associate a partial fraction A/(ax + b), where A is a constant ≠ 0.

Example 3.21

$$\frac{x+4}{(x+7)(2x-1)} = \frac{A}{x+7} + \frac{B}{2x-1}$$

(2)   Linear factors some of which are repeated.

If a linear factor ax + b occurs p times as a factor of the denominator of the given fraction, then corresponding to this factor associate the p partial fractions

$$\frac{A_1}{ax+b} + \frac{A_2}{(ax+b)^2} + \cdots + \frac{A_p}{(ax+b)^p}$$

where $A_1, A_2, ..., A_p$ are constants and $A_p \neq 0$.

Example 3.22

$$\frac{3x-1}{(x+4)^2} = \frac{A}{x+4} + \frac{B}{(x+4)^2}$$

(3)   Quadratic factors none of which are repeated.

If a quadratic factor $ax^2 + bx + c$ occurs once as a factor of the denominator of the given fraction, then corresponding to this factor associate a partial fraction

$$\frac{Ax+B}{ax^2+bx+c}$$

where A and B are constants which are not both zero.

Note. It is assumed that $ax^2 + bx + c$ cannot be factored into two real linear factors with integer coefficients.

Example 3.23

$$\frac{x^2-3}{(x-2)(x^2+4)} = \frac{A}{x-2} + \frac{Bx+C}{x^2+4}$$

(4)   Quadratic factors some of which are repeated.

If an irreducible quadratic factor $ax^2 + bx + c$ occurs p times as a factor of the denominator of the given fraction, then corresponding to this factor associate the p partial fractions

$$\frac{A_1x+B_1}{ax^2+bx+c} \quad , \quad \frac{A_2x+B_2}{(ax^2+bx+c)^2} \quad , \quad , \quad \frac{A_px+B_p}{(ax^2+bx+c)^p}$$

where A_1, B_1, A_2, B_2, ..., A_p, B_p are constants and A_p, B_p are not both zero.

Example 3.24

$$\frac{x^2-4x+1}{\left(x^2+1\right)^2\left(x^2+x+1\right)}=\frac{Ax+B}{x^2+1}+\frac{Cx+D}{\left(x^2+1\right)^2}+\frac{Ex+F}{x^2+x+1}$$

Finding the Partial Fraction Decomposition

Once the form of the partial fraction decomposition of a rational fraction has been determined, the next step is to find the system of equations to be solved to get the values of the constants needed in the partial fraction decomposition. Although the system of equations usually involves more than three equations, it is often quite easy to determine the value of one or two variables or relationships among the variables that allows the system to be reduced to a size small enough to be solved conveniently by any method.

Example 3.25. Find the partial fraction decomposition of

$$\frac{3x^2+3x+7}{\left(x-2\right)^2\left(x^2+1\right)}$$

Using Rules (2) and (3) of the previous section, the form of the decomposition is:

$$\frac{3x^2+3x+7}{\left(x-2\right)^2\left(x^2+1\right)}=\frac{A}{x-2}+\frac{B}{\left(x-2\right)^2}+\frac{Cx+D}{x^2+1}$$

$$\frac{3x^2+3x+7}{\left(x-2\right)^2\left(x^2+1\right)}=\frac{A\left(x-2\right)\left(x^2+1\right)+B\left(x^2+1\right)+\left(Cx+D\right)\left(x-2\right)^2}{\left(x-2\right)^2\left(x^2+1\right)}$$

$$3x^2 + 3x + 7 = Ax^3 - 2Ax^2 + Ax - 2A + Bx^2 + B + Cx^3 - 4Cx^2 + Dx^2 + 4Cx - 4Dx + 4D$$

$3x^2 + 3x + 7 = (A + C)x^3 + (-2A + B - 4C + D)x^2 + (A + 4C - 4D)x +$
$(-2A + B + 4D)$

Equating the coefficients of the corresponding terms in the two polynomials and setting the others equal to 0, we get the system of equations to solve:

$$A + C = 0$$

$$-2A + B - 4C + D = 3$$

$$A + 4C - 4D = 3$$

$$-2A + B + 4D = 7$$

Solving the system, we get A = -1, B = 5, C = 1, and D = 0. Thus, the partial fraction decomposition is:

$$\frac{3x^2+3x+7}{(x-2)^2(x^2+1)} = \frac{-1}{x-2} + \frac{5}{(x-2)^2} + \frac{x}{x^2+1}$$

Chapter 4
LINEAR EQUATIONS

IN THIS CHAPTER:

✔ *Equations*
✔ *Linear Equations*
✔ *Equations of Lines*
✔ *Simultaneous Linear Equations*
✔ *Inequalities*
✔ *Determinants and Systems of Linear Equations*

Equations

An **equation** is a statement of equality between two expressions called members. An equation that is true for only certain values of the variables (sometimes called unknowns) involved is called a **conditional equation** or simply an equation. An equation that is true for all permissible values of the variables (or unknowns) involved is called an **identity.** By permissible values, we mean the values for which the members are defined.

Example 4.1 $x + 5 = 8$ is true only for $x = 3$; it is a conditional equation.

Example 4.2 $x^2 - y^2 = (x - y)(x + y)$ is true for all values of x and y; it is an identity.

The solutions of a conditional equation are those values of the unknowns that make both members equal. These solutions are said to satisfy the equation. If only one unknown is involved, the solutions are also called **roots**. To solve an equation means to find all of the solutions.

Thus, $x = 2$ is a solution or root of $2x + 3 = 7$, since if we substitute $x = 2$ into the equation, we obtain $2(2) + 3 = 7$ and both members are equal, i.e., the equation is satisfied. Similarly, three (of the many) solutions of $2x + y = 4$ are: $x = 0$, $y = 4$; $x = 1$, $y = 2$; $x = 5$, $y = -6$.

Operations Used in Transforming Equations

- If equals are added to equals, the results are equal.
  Thus, if $x - y = z$, we may add y to both members and obtain $x = y + z$.

- If equals are subtracted from equals, the results are equal.
  Thus, if $x + 2 = 5$, we may subtract 2 from both members to obtain $x = 3$.

- If equals are multiplied by equals, the results are equal.
  Thus, if both members of $(1/4)y = 2x^2$ are multiplied by 4, the result is $y = 8x^2$.

- If equals are divided by equals, the results are equal, provided there is no division by zero.
  Thus, if $-4x = -12$, we may divide both members by -4 to obtain $x = 3$.

- The same powers of equals are equal.

  Thus, if $T = 2\pi \sqrt{(1/g)}$, then $T^2 = (2\pi \sqrt{(1/g)})^2 = 4\pi^2 1/g$.

- The same roots of equals are equal. Thus:

$$\text{if } r^3 = \frac{3V}{4\pi}, \text{ then } r = \sqrt[3]{\frac{3V}{4\pi}}$$

- Reciprocals of equals are equal provided the reciprocal of zero does not occur. Thus, if $1/x = 1/3$, then $x = 3$.

Formulas

A **formula** is an equation which expresses a general fact, rule, or principle.

For example, in geometry, the formula $A = \pi r^2$ gives the area A of a circle in terms of its radius r.

In physics, the formula $s = (1/2)gt^2$, where g is approximately 32.2 ft/s^2, gives the relation between the distance s, in feet, which an object will fall freely from rest during a time t, in seconds.

To solve a formula for one of the variables involved is to perform the same operations on both members of the formula until the desired variable appears on one side of the equation but not on the other side.

Polynomial Equations

A **monomial** in a number of unknowns x, y, z, ... has the form $ax^p y^q z^r$ ••• where the exponents p, q, r, ... are either positive integers or zero and the coefficient a is independent of the unknowns. The sum of the exponents $p + q + r + $ ••• is called the degree of the term in the unknowns x, y, z, ...

Examples 4.3 $3x^2z^3$, $(1/2)x^4$, 6 are monomials.
$3x^2z^3$ is of degree 2 in x, 3 in z, and 5 in x and z.

You Need to Know

When reference is made to degree without specifying the unknowns considered, the degree in all unknowns is implied.

A polynomial in various unknowns consist of terms, each of which is rational and integral. The degree of such a polynomial is defined as the degree of the terms of highest degree.

Examples 4.4 $3x^3y^4z + xy^2z^5 - 8x + 3$ is a polynomial of degree 3 in x, 4 in y, 5 in z, 7 in x, 7 in y and z, 6 in x and z, and 8 in x, y, and z.

A **polynomial equation** of degree n in the unknown x may be written

$$a_0x^n + a_1x^{n-1} + a_2x^{n-2} + \cdots + a_{n-1}x + a_n = 0; \quad a_0 \neq 0$$

where a_0, a_1, ... , a_n are given constants and n is a positive integer.
As special cases, we see that

$$a_0x + a_1 = 0 \qquad \text{or} \qquad ax + b = 0$$
is of degree 1 (linear equation).

$$a_0x^2 + a_1x + a_2 = 0 \qquad \text{or} \qquad ax^2 + bx + c = 0$$
is of degree 2 (quadratic equation).

$$a_0x^3 + a_1x^2 + a_2x + a_3 = 0$$
is of degree 3 (cubic equation).

Linear Equations

A **linear equation** in one variable has the form $ax + b = 0$, where $a \neq 0$ and b are constants. The solution of this equation is given by $x = -b/a$.

When a linear equation is not in the form $ax + b = 0$, we simplify the equation by multiplying each term by the LCD for all fractions in the equation, removing any parentheses, or combining like terms. In some equations, we do more than one of the procedures.

Example 4.5 Solve the equation $x + 8 - 2(x + 1) = 3x - 6$ for x.

$x + 8 - 2(x + 1) = 3x - 6$ First we remove the parentheses.

$x + 8 - 2x - 2 = 3x - 6$ We now combine like terms.

$- x + 6 = 3x - 6$ We simplify the above equation.

$- x + 6 - 3x = 3x - 6 - 3x$ Now get the variable terms on one side of the equation to isolate the variable term on one side of the equation by itself.

$- 4x + 6 = - 6$ Simplifying the above equation.

$- 4x + 6 - 6 = - 6 - 6$ Now we subtract 6 from each side of the equation to get the variable term on one side of the equation by itself.

$- 4x = -12$ Simplifying the above equation.

$$\frac{-4x}{-4} = \frac{-12}{-4}$$

Finally, we divide each side by the coefficient of the variable, which is -4.

$x = 3$

Now we check the solution in the original equation.

Check:

$3 + 8 - 2(3 + 1) \overset{?}{=} 3(3) - 6$

The question mark indicates that we don't know for sure that the two quantities are equal.

$11 - 2(4) \overset{?}{=} 9 - 6$

$11 - 8 \overset{?}{=} 3$

$3 = 3$

The solution checks.

Word Problems

In solving a word problem, the first step is to decide what is to be found. The next step is to translate the conditions stated in the problem into an equation or to state a formula that expresses the conditions of the problem. The solution of the equation is the next step.

Example 4.6 If the perimeter of a rectangle is 68 meters and the length is 14 meters more than the width, what are the dimensions of the rectangle?

Let w = the number of meters in the width and w + 14 = the number of meters in the length.

$2[(w + 14) + w] = 68$

$2w + 28 + 2w = 68$

$4w + 28 = 68$

$4w = 40$

$w = 10$

$w + 14 = 24$

The rectangle is 24 meters long by 10 meters wide.

Example 4.7 How many liters of pure alcohol must be added to 15 liters of a 60% alcohol solution to obtain an 80% alcohol solution?

Let n = the number of liters of pure alcohol to be added.

$n + 0.60(15) = 0.80(n + 15)$ (The sum of the amounts of alcohol in each quantity is equal to the amount of alcohol in the mixture.)

$n + 9 = 0.8n + 12$

$0.2n = 3$

$n = 15$

Fifteen liters of pure alcohol must be added.

Equations of Lines

Slope of a Line

The equation ax + by = c, where not both a and b are 0 and a, b, and c are real numbers is the standard (or general) form of the equation of a line. The **slope**, m, of a line through two points (x_1, y_1) and (x_2, y_2) is defined as the ratio of the change in y compared to the change in x or:

$$m = \frac{\text{change in y}}{\text{change in x}} = \frac{y_2 - y_1}{x_2 - x_1}$$

with $x_1 \neq x_2$.

Example 4.8 What is the slope of the line 3x - 4y = 12?

First we need to find two points that satisfy the equation of the line 3x - 4y = 12. If x = 0, then 3(0) - 4y = 12 and y = -3. Thus, one point is (0, -3). If x = -4, then 3(-4) - 4y = 12 and y = -6. So, (-4, -6) is another point on the line.

$$m = \frac{y_2 - y_1}{x_2 - x_1} = \frac{-3 - (-6)}{0 - (-4)} = \frac{3}{4}$$

The slope of the line 3x - 4y = 12 is 3/4.

- A positive slope means that, as x increases, so does y.

- A negative slope means that, as x increases, y decreases.

- A horizontal line y = k, where k is a constant, has zero slope.

- A vertical line x = k, where k is a constant, does not have a slope, that is, the slope is not defined.

Slope-Intercept Form of Equation of a Line

If a line has slope m and y intercept (0, b), then for any point (x, y) where x ≠ 0, on the line, we have

$$m = \frac{y-b}{x-0} \text{ and } y = mx+b$$

The slope-intercept form of the equation of a line with slope m and y intercept b is y = mx + b.

Example 4.9 Find the equation of the line with slope -4 and y intercept 6.

The slope of the line is -4, so m = -4 and the y intercept is 6, so b = 6. Substituting into y = mx + b, we get y = -4x + 6 for the equation of the line.

Slope-Point Form of Equation of a Line

If a line has slope m and goes through a point (x_1, y_1), then for any other point (x, y) on the line, we have m = $(y - y_1)/(x - x_1)$ and $y - y_1 = m(x - x_1)$. The slope-point form of the equation of a line is $y - y_1 = m(x - x_1)$

Example 4.10 Write the equation of the line passing through the point (1, -2) and having slope -2/3.

Since (x_1, y_1) = (1, -2) and m = -2/3, we substitute into $y - y_1 = m(x - x_1)$ to get y + 2 = -2/3(x - 1). Simplifying, we get 3(y + 2) = -2(x - 1), and finally 2x + 3y = -4.

Two-Point Form of Equation of a Line

If a line goes through the points (x_1, y_1) and (x_2, y_2), it has a slope m = $(y_2 - y_1)/(x_2 - x_1)$ if $x_2 \neq x_1$. Substituting into the equation $y - y_1 = m(x - x_1)$, we get

$$y - y_1 = \frac{y_2 - y_1}{x_2 - x_1}(x - x_1)$$

The two-point form of the equation of a line is

$$y - y_1 = \frac{y_2 - y_1}{x_2 - x_1}(x - x_1)$$

if $x_2 \neq x_1$.

- If $x_2 = x_1$, we get the vertical line $x = x_1$.
- If $y_2 = y_1$, we get the horizontal line $y = y_1$.

Example 4.11 Write the equation of the line passing through $(3, 6)$ and $(-4, 4)$.

Let $(x_1, y_1) = (3, 6)$ and $(x_2, y_2) = (-4, 4)$ and substitute into

$$y - y_1 = \frac{y_2 - y_1}{x_2 - x_1}(x - x_1)$$

$$y - 6 = \frac{4 - 6}{-4 - 3}(x - 3)$$

$-7(y - 6) = -2(x - 3)$

$-7y + 42 = -2x + 6$

$2x - 7y = -36$

The equation of the line through the points $(3, 6)$ and $(-4, 4)$ is $2x - 7y = -36$.

Intercept Form of Equation of a Line

If a line has x intercept a and y intercept b, it goes through the points (a, 0) and (0, b). The equation of the line is

$$y - b = \frac{0 - b}{a - 0}(x - 0)$$

if a ≠ 0, which simplifies to bx + ay = ab. If both a and b are non-zero, we get

$$\frac{x}{a} + \frac{y}{b} = 1$$

If a line has x intercept a and y intercept b and both a and b are non-zero, the equation of the line is

$$\frac{x}{a} + \frac{y}{b} = 1$$

Example 4.12 Find the intercepts of the line 4x - 3y = 12.

We divide the equation 4x - 3y = 12 by 12 to get

$$\frac{x}{3} + \frac{y}{-4} = 1$$

The x intercept is 3 and the y intercept is -4 for the line 4x - 3y = 12.

Simultaneous Linear Equations

Systems of Two Linear Equations

A linear equation in two variables x and y is of the form $ax + by = c$ where a, b, c are constants and a, b are not both zero. If we consider two such equations

$$a_1x + b_1y = c_1$$
$$a_2x + b_2y = c_2$$

we say that we have two simultaneous linear equations in two unknowns or a system of two linear equations in two unknowns. A pair of values for x and y, (x, y), which satisfies both equations is called a **simultaneous solution** of the given equation. Thus, the simultaneous solution of $x + y = 7$ and $x - y = 3$ is (5, 2).

Three methods of solving such systems of linear equations are illustrated here.

- *Solution by addition or subtraction.* If necessary, multiply the given equations by such numbers as will make the coefficients of one unknown in the resulting equations numerically equal. If the signs of the equal coefficients are unlike, add the resulting equations; if like, subtract them.

  For example, consider the two equations:

  (1) $2x - y = 4$
  (2) $x + 2y = -3$

  To eliminate y, multiply (1) by 2 and add to (2) to obtain

  2 (1): $4x - 2y = 8$
    (2): $x + 2y = -3$
  _____
  Addition: $5x = 5$ or $x = 1$

Substitute x = 1 in (1) and obtain 2 - y = 4 or y = -2.
Thus, the simultaneous solution of (1) and (2) is (1, 2).

Check: Put x = 1, y = -2 in (2) and obtain 1 + 2(-2)
? -3 or -3 = -3.

- *Solution by substitution.* Find the value of one unknown in terms of the other inknown in either of the given equations and substitute this value in the other equation.

  For example, consider the system (1) and (2) above. From (1), obtain y = 2x - 4 and substitute this value into (2) to get x + 2(2x - 4) = -3 which reduces to x = 1. Then put x = 1 into either (1) or (2) and obtain y = -2. The solution is (1, -2).

- *Graphical solution.* Graph both equations, obtaining two straight lines. The simultaneous solution is given by the coordinates (x, y) of the point of intersection of these lines. Figure 4-1 shows that the simultaneous solution of (1) 2x - y = 4 and (2) x + 2y = -3 is x = 1, y = -2, also written (1, -2).

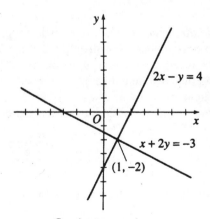

Consistent equations

(1) $2x - y = 4$
(2) $x + 2y = -3$

Figure 4-1

If the lines are parallel, the equations are **inconsistent** and have no simultaneous solution.

For example, (3) $x + y = 2$ and (4) $2x + 2y = 8$ are inconsistent, as indicated by Figure 4-2.

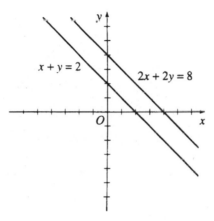

Inconsistent equations

(3) $x + y = 2$
(4) $2x + 2y = 8$

Figure 4-2

Dependent equations are represented by the same lines. Thus every point on the line represents a solution and, since there are an infinite number of points, there are an infinite number of simultaneous solutions. For example, (5) $x + y = 1$ and (6) $4x + 4y = 4$ are dependent equations as indicated in Figure 4-3.

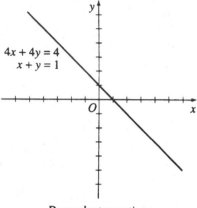

Dependent equations

(5) $x + y = 1$

(6) $4x + 4y = 4$

Figure 4-3

Systems of Three Linear Equations

A system of three linear equations in three variables is solved by eliminating one unknown from any two of the equations and then eliminating the same unknown from any other pair of equations.

Linear equations in three variables represent planes and can result in two or more parallel planes, which are thus inconsistent and have no solution. The three planes can coincide or all three can intersect in a common line and be dependent. The three planes can intersect in a single point, like the ceiling and two walls forming a corner in a room, and are consistent.

Linear equations in three variables x, y, and z are of the form ax + by + cz = d, where a, b, c, and d are real numbers and not all three of a, b, and c are zero. If we consider three such equations

$$a_1x + b_1y + c_1z = d_1$$
$$a_2x + b_2y + c_2z = d_2$$

$$a_3x + b_3y + c_3z = d_3$$

and find a value (x, y, z) that satisfies all three equations, we say we have a simultaneous solution to the system of equations.

Example 4.13 Solve the system of equations

(1) $2x + 5y + 4z = 4$
(2) $x + 4y + 3z = 1$
(3) $x - 3y - 2z = 5$

First we will eliminate x from (1) and (2) and from (2) and (3).

$$
\begin{array}{ll}
\quad\; 2x + 5y + 4z = 4 & \quad\quad x + 4y + 3z = 1 \\
- 2x - \;\; 8y - \;6z = -2 & \quad - x + 3y + 2z = -5 \\
\hline
\end{array}
$$

(4) $-3y - 2z = 2$ (5) $7y + 5z = -4$

Now we eliminate z from equations (4) and (5).

$$
\begin{array}{l}
-15y - 10z = 10 \\
\;\,14y + 10z = -8 \\
\hline
\end{array}
$$

(6) $-y = 2$

We solve (6) and get $y = -2$. Substituting into (4) or (5), we solve for z.

(4) $-3(-2) - 2z = 2$
                       $+6 - 2z = 2$
                       $-2z = -4$
                       $z = 2$

Substituting into (1), (2), or (3), we solve for x.

(1)
$$2x + 5(-2) + 4(2) = 4$$
$$2x - 10 + 8 = 4$$
$$2x - 2 = 4$$
$$2x = 6$$
$$x = 3$$

The solution to the system of equations is (3, -2, 2).

We check the solution by substituting the point (3, -2, 2) into equations (1), (2), and (3).

(1) $2(3) + 5(-2) + 4(2) \overset{?}{=} 4$

   $6 - 10 + 8 \overset{?}{=} 4$

   $4 = 4$

(2) $3 + 4(-2) + 3(2) \overset{?}{=} 1$

   $3 - 8 + 6 \overset{?}{=} 1$

   $1 = 1$

(3) $3 - 3(-2) - 2(2) \overset{?}{=} 5$

   $3 + 6 - 4 \overset{?}{=} 5$

   $5 = 5$

Thus, (3, -2, 2) checks in each of the given equations and is the answer to the problem.

Inequalities

An **inequality** is a statement that one real quantity or expression is greater or less than another real quantity or expression. The following indicate the meaning of inequality signs:

(1) $a > b$ means "a is greater than b" (or a - b is a positive number).

(2) $a < b$ means "a is less than b" (or a - b is a negative number).

(3) a ≥ b means "a is greater than or equal to b."

(4) a ≤ b means "a is less than or equal to b."

(5) 0 < a < 2 means "a is greater than zero but less than 2."

(6) -2 ≤ x < 2 means "x is greater than or equal to -2 but less than 2."

An **absolute inequality** is true for all real values of the letters involved. For example, $(a - b)^2 > -1$ holds for all real values of a and b, since the square of any real number is positive or zero.

A **conditional inequality** holds only for particular values of the letters involved. Thus $x - 5 > 3$ is true only when x is greater than 8.

The inequalities $a > b$ and $c > d$ have the same sense. The inequalities $a > b$ and $x < y$ have opposite sense.

Principles of Inequalities

(1) The sense of an inequality is unchanged if each side is increased or decreased by the same real number. It follows that any term may be transposed from one side of an inequality to the other, provided the sign of the term is changed.

Thus, if $a > b$, then $a+c > b+c$, and $a-c > b-c$, and $a-b > 0$.

(2) The sense of an inequality is unchanged if each side is multiplied or divided by the same positive number.

Thus, if $a > b$ and $k > 0$, then

$$ka > kb \qquad \text{and} \qquad \frac{a}{k} > \frac{b}{k}$$

(3) The sense of an inequality is reversed if each side is multiplied or divided by the same negative number.

Thus, if $a > b$ and $k < 0$, then

$$ka < kb \qquad \text{and} \qquad \frac{a}{k} < \frac{b}{k}$$

(4) If $a > b$ and a, b, n are positive, then $a^n > b^n$ but $a^{-n} < b^{-n}$.

Examples 4.14

(a) $5 > 4$; then $5^3 > 4^3$ or $125 > 64$, but

$$5^{-3} < 4^{-3} \quad \text{or} \quad \frac{1}{125} < \frac{1}{64}$$

(b) $16 > 9$; then $16^{1/2} > 9^{1/2}$ or $4 > 3$, but

$$16^{-1/2} < 9^{-1/2} \quad \text{or} \quad \frac{1}{4} < \frac{1}{3}$$

(5) If $a > b$ and $c > d$, then $(a + c) > (b + d)$.

(6) If $a > b > 0$ and $c > d > 0$, then $ac > bd$.

Absolute Value Inequalities

The **absolute value** of a quantity represents the distance that the value of the expression is from zero on a number line. So $|x - a| = b$, where $b > 0$, says that the quantity $x - a$ is b units from 0, $x - a$ is b units to the right of 0, or $x - a$ is b units to the left of 0. When we say $|x - a| > b$, $b > 0$, then $x - a$ is at a distance from 0 that is greater than b. Thus, $x - a > b$ or $x - a < -b$. Similarly, if $|x - a| < b$, $b > 0$, then $x - a$ is at a distance from 0 that is less than b. Hence, x - a is between b units below 0, -b, and b units above 0.

Examples 4.15 Solve each of these inequalities for x.

(a) $|x- 3| > 4$ (b) $|x+4| < 7$ (c) $|x-5| < -3$ (d) $|x-5| > -5$

(a) $|x-3| > 4$, then x-3 > 4 or x- 3 < -4. Thus, x > 7 or x < -1. The solution set is $(-\infty, -1) \cup \times z (7,\infty)$, (where $\cup \times z$ represents the union of the two intervals).

(b) $|x + 4| < 7$ then - 7 < x + 4 < 7. Thus, - 11 < x < 3. The solution interval is (-11, 3).

(c) $|x - 5| < -3$. Since the absolute value of a number is always greater than or equal to zero, there are no values for which the absolute value will be less than -3. Thus, there is no solution and we may write Ø for the solution interval.

(d) $|x + 3| > -5$. Since the absolute value of a number is always at least zero, it is always greater than -5. Thus the solution is all real numbers, and for the solution interval we write $(-\infty, \infty)$.

Higher Degree Inequalities

Solving higher degree inequalities is similar to solving higher degree equations: we must always compare the expression to zero. If f(x) > 0, then we are interested in the values of x that will produce a product and/or quotient of factors that is positive, while if f(x) < 0, we wish to find the values of x that will produce a product and/or quotient that is negative.

If f(x) is a quadratic expression, we have just two factors to consider, and we can do this by examining cases based on the possible signs of the two factors that will produce the desired sign for the expression. When the number of factors in f(x) increases by one, the number of cases to consider doubles. Thus, for an expression with 2 factors there are 4 cases, with 3 factors there are 8 cases, and with 4 factors there are 16 cases. In each instance, half of the cases will produce a positive expression and half a negative one. Thus, the case procedure gets to be a very long one quite quickly. An alternative procedure to the case method is the sign chart.

Example 4.16 Solve the inequality $x^2 + 15 < 8x$.

The inequality $x^2 + 8x < 15$ is equivalent to $x^2 - 8x + 15 < 0$ and to $(x - 3)(x - 5) < 0$ and is true when the product of $x - 3$ and $x - 5$ is negative. The critical values of the product are the values that make these factors 0, because they represent where the product may change signs.

The critical values of x, 3 and 5, are placed on a number line and divide it into three intervals. We need to find the sign of the product of $x - 3$ and $x - 5$ on each of these intervals to find the solution (see Figure 4-4).

Figure 4-4

Vertical lines are drawn through each critical value. A dashed line indicates that the critical value is not in the solution, and a solid line indicates that the critical value is in the solution.

The signs above the number line are the signs for the factors and are found by selecting an arbitrary value in the interval as a test value and determining whether each factor is positive or negative for the test value. For the interval to the left of 3, we choose a test value of 1 and substitute it into $x - 3$ and see that the value is -2, so we record a – sign, and for $x - 5$ the value is -4 and again we record a – sign. For the interval between 3 and 5 we choose any value, such as 3.5, and determine that x - 3 is positive and x - 5 is negative. Finally, for the interval to the right of 5, we choose a value of 12 and see that both x - 3 and x - 5 are positive. The sign for the problem, written below the line, in each interval is determined by the signs of the factors in that interval. If an even number of factors in a product or quotient are negative, the product or quotient is positive. If an odd number of factors are negative, the product or quotient is negative.

We select the intervals that satisfy our problem (x - 3)(x - 5) < 0, so we select the intervals that are negative in the sign chart. In the interval between 3 and 5 the problem is negative (see Figure 4-4), so the solution is the interval (3, 5). The parentheses mean that the 3 and 5 are not included in the interval, and we know this since the boundary lines are dashed. If they had been in the solution, we would have used a bracket instead of a parenthesis at the end of the interval next to the 3.

The solution for $x^2 + 15 < 8x$ is the interval (3, 5).

Example 4.17 Solve the inequality

$$\frac{x-3}{x(x+4)} \geq 0$$

The inequality is compared to 0 and the numerator and denominator are factored, so we can see that the critical values for the problem are the solution of x = 0, x - 3 = 0, and x + 4 = 0. Thus, the critical values are x = 0, x = 3 and x = -4. Since there are three critical values, the number line is divided into four distinct intervals, as shown in Figure 4-5.

$x - 3$	−	−	−	+
x	−	−	+	+
$x + 4$	−	+	+	+
Problem	−	+	−	+
	−4	0	3	

Figure 4-5

The signs above the line are the signs of each factor in each interval. The sign below is the sign for the problem, and it is + when an even number of factors are negative and − when an odd number of factors are negative. Since the problem uses the ≥ sign, values that make the numerator zero are solutions, so a solid line is drawn through 3. Since 0 and −4 make the denominator of the fraction 0, they are not solutions and dashed lines are drawn through 0 and −4 (see Figure 4-5).

Since the problem indicates that a positive or zero value is wanted, we want the regions with a + sign in the sign chart. Thus, the solutions are the intervals, (−4, 0) and [3, ∞), and the solution is written (−4, 0) ∪ [3, ∞). The ∪ indicates that we want the union of the two intervals. Note that the bracket, [, is used because the critical value 3 is in the solution and a parenthesis,), is always used for the infinite, ∞, side of an interval.

Linear Inequalities in Two Variables

The solution of linear inequalities in two variables x and y consists of all points (x,y) that satisfy the inequality. Since a linear equation represents a line, a linear inequality is the points on one side of a line. The points on the line are included when the sign ≥ or ≤ is used in the statement of the inequality. The solutions of linear inequalities are usually found by graphical methods.

Example 4.18 Find the solution for 2x − y ≤ 3.

We graph the line related to the inequality 2x − y ≤ 3, which is 2x − y = 3. Since the symbol ≤ is used, the line is part of the solution, and a solid line is used to indicate this (see Figure 4-6).

If the line is not part of the solution, we use a dashed line to indicate that fact. We shade the region on the side of the line where the points are solutions of the inequality. The solution region is determined by selecting a test point that is not on the line. If the test point satisfies the inequality, then all points on that side of the line are in the solution. If the test point does not satisfy the inequality, no points on that side of the

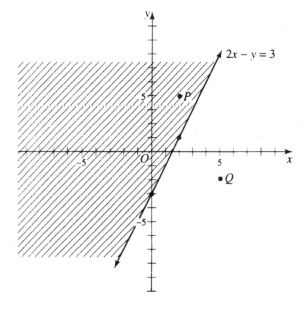

Figure 4-6

line are in the solution. Hence the solution points are on the opposite side of the line from the test point.

The point P(2, 4) is not on the line 2x - y = 3, so it can be used as a test point. When we substitute (2, 4) into the inequality 2x - y ≤ 3, we get 2(2) - 4 ≤ 3, which is true, since 0 ≤ 3. We shade on the side of the line that contains the test point (2, 4) to indicate the solution region. If we had selected Q(5, -2) and substituted into 2x - y ≤ 3, we would have obtained 12 ≤ 3, which is false, and would have shaded on the opposite side of the line from Q. This is the same region we found using the test point P.

The solution for 2x - y ≤ 3 is shown in Figure 4-6 and consists of the shaded region and the line.

Systems of Linear Inequalities

If we have two or more linear inequalities in two variables, we say we have a system of linear inequalities and the solution of the system is the inter-section, or common region, of the solution regions for the inequalities.

> A system with two inequalities whose related equations intersect always has a solution region. If the related equations are parallel, the system may or may not have a solution. Systems with three or more inequalities may or may not have a solution.

Example 4.19 Solve the system of inequalities $2x + y > 3$ and $x - 2y \leq$ -1.

We graph the related equations $2x + y = 3$ and $x - 2y = -1$ on the same set of axes. The line $2x + y = 3$ is dashed, since it is not included in $2x + y > 3$, but the line $x - 2y = -1$ is solid, since it is included in $x - 2y \leq$ -1.

Now we select a test such as $(0, 5)$ that is not on either line, deter-mine which side of each line to shade and shade only the common region. Since $2(0) + 5 > 3$ is true, the solution region is to the right and above the line $2x + y = 3$. Since $0 - 2(5) \leq -1$ is true, the solution region is to the left and above the line $x - 2y = -1$.

The solution region of $2x + y > 3$ and $x - 2y \leq 1$ is the shaded region of Figure 4-7, which includes the part of the solid line bordering the shaded region.

Determinants and Systems of Linear Equations

Determinants of Second Order

The symbol

$$\begin{vmatrix} a_1 & b_1 \\ a_2 & b_2 \end{vmatrix}$$

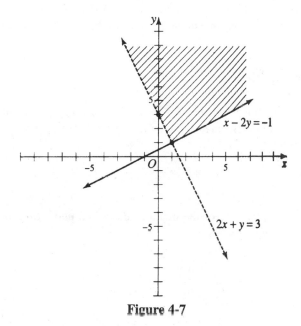

Figure 4-7

consisting of the four numbers a_1, b_1, a_2, b_2 arranged in two rows and two columns, is called a **determinant of second order** or **determinant of order two**. The four numbers are called **elements** of the determinant. By definition,

$$\begin{vmatrix} a_1 & b_1 \\ a_2 & b_2 \end{vmatrix} = a_1 b_2 - b_1 a_2$$

Thus,

$$\begin{vmatrix} 2 & 3 \\ -1 & -2 \end{vmatrix} = (2)(-2) - (3)(-1) = -4 + 3 = -1$$

Here the elements 2 and 3 are in the first row, and the elements - 1 and - 2 are in the second row. Elements 2 and - 1 are in the first column, and elements 3 and -2 are in the second column. A determinant is a number. A determinant of order one is the number itself.

Cramer's Rule

Systems of two linear equations in two unknowns may be solved by use of second-order determinants. Given the system of equations

$$a_1x + b_1y = c_1 \qquad\qquad (4.1)$$
$$a_2x + b_2y = c_2$$

it is possible to use any of the methods described under *Simultaneous Linear Equations* to obtain

$$x = \frac{c_1b_2 - b_1c_2}{a_1b_2 - b_1a_2}, \quad y = \frac{a_1c_2 - c_1a_2}{a_1b_2 - b_1a_2} \quad \left(a_1b_2 - b_1a_2 \neq 0\right)$$

These values for x and y may be written in terms of second-order determinants as follows:

$$x = \frac{\begin{vmatrix} c_1 & b_1 \\ c_2 & b_2 \end{vmatrix}}{\begin{vmatrix} a_1 & b_1 \\ a_2 & b_2 \end{vmatrix}}, \quad y = \frac{\begin{vmatrix} a_1 & c_1 \\ a_2 & c_2 \end{vmatrix}}{\begin{vmatrix} a_1 & b_1 \\ a_2 & b_2 \end{vmatrix}} \qquad\qquad (4.2)$$

The form involving determinants is easy to remember if one keeps in mind the following:

(a) The denominators in (4.2) are given by the determinant
$$\begin{vmatrix} a_1 & b_1 \\ a_2 & b_2 \end{vmatrix}$$

in which the elements are the coefficients of x and y arranged as in the given equations of (4.1). This determinant, usually denoted by D, is called the **determinant of the coefficients**.

(b)   The numerator in the solution for either unknown is the same as the determinant of the coefficients D with the exception that the column of coefficients of the unknown to be determined is replaced by the column of constants on the right side of (4.1). When the column of coefficients for the variable x in determinant D is replaced with the column of constants, we call the new determinant D_x. When the column of y coefficients in determinant D is replaced with the column of constants, we call the new determinant D_y.

Example 4.20 Solve the system

2x + 3y = 8
x - 2y = - 3

The denominator for both x and y is $D = \begin{vmatrix} 2 & 3 \\ 1 & -2 \end{vmatrix} = 2(-2) - 3(1) = -7$

$$D_x = \begin{vmatrix} 8 & 3 \\ -3 & -2 \end{vmatrix} = 8(-2) - 3(-3) = -7$$

$$D_y = \begin{vmatrix} 2 & 8 \\ 1 & -3 \end{vmatrix} = 2(-3) - 8(1) = -14$$

$$x = \frac{D_x}{D} = \frac{-7}{-7} = 1, \quad y = \frac{D_y}{D} = \frac{-14}{-7} = 2$$

Thus, the solution of the system is (1, 2).

You Need to Know

The method of solution of linear equations by determinants is called **Cramer's Rule**. If the determinant D = 0, then Cramer's Rule can not be used to solve the system.

Determinants of Third Order

The symbol

$$\begin{vmatrix} a_1 & b_1 & c_1 \\ a_2 & b_2 & c_2 \\ a_3 & b_3 & c_3 \end{vmatrix}$$

consisting of nine numbers arranged in three rows and three columns is called a **determinant of third order**. By definition, the value of this determinant is given by

$$a_1b_2c_3 + b_1c_2a_3 + c_1a_2b_3 - c_1b_2a_3 - a_1c_2b_3 - b_1a_2c_3$$

Cramer's Rule for linear equations in 3 unknowns is a method of solving the following equations for x, y, z

$$a_1x + b_1y + c_1z = d_1$$
$$a_2x + b_2y + c_2z = d_2 \tag{4.3}$$
$$a_3x + b_3y + c_3z = d_3$$

It is an extension of Cramer's Rule for linear equations in two unknowns. We can solve the equations in (4.3) to yield

$$x = \frac{d_1 b_2 c_3 + c_1 d_2 b_3 + b_1 c_2 d_3 - c_1 b_2 d_3 - b_1 d_2 c_3 - d_1 c_2 b_3}{a_1 b_2 c_3 + b_1 c_2 a_3 + c_1 a_2 b_3 - c_1 b_2 a_3 - b_1 a_2 c_3 - a_1 c_2 b_3}$$

$$y = \frac{a_1 d_2 c_3 + c_1 a_2 d_3 + d_1 c_2 a_3 - c_1 d_2 a_3 - d_1 a_2 c_3 - a_1 c_2 d_3}{a_1 b_2 c_3 + b_1 c_2 a_3 + c_1 a_2 b_3 - c_1 b_2 a_3 - b_1 a_2 c_3 - a_1 c_2 b_3}$$

$$z = \frac{a_1 b_2 d_3 + d_1 a_2 b_3 + b_1 d_2 a_3 - d_1 b_2 a_3 - b_1 a_2 d_3 - a_1 d_2 b_2}{a_1 b_2 c_3 + b_1 c_2 a_3 + c_1 a_2 b_3 - c_1 b_2 a_3 - b_1 a_2 c_3 - a_1 c_2 b_3}$$

These may be written in terms of determinants as follows

$$D = \begin{vmatrix} a_1 & b_1 & c_1 \\ a_2 & b_2 & c_2 \\ a_3 & b_3 & c_3 \end{vmatrix} \quad D_x = \begin{vmatrix} d_1 & b_1 & c_1 \\ d_2 & b_2 & c_2 \\ d_3 & b_3 & c_3 \end{vmatrix} \quad D_y = \begin{vmatrix} a_1 & d_1 & c_1 \\ a_2 & d_2 & c_2 \\ a_3 & d_3 & c_3 \end{vmatrix} \quad D_z = \begin{vmatrix} a_1 & b_1 & d_1 \\ a_2 & b_2 & d_2 \\ a_3 & b_3 & d_3 \end{vmatrix}$$

D is the determinant of the coefficients of x, y, z in (4.3) and is assumed not equal to zero. If D is zero, Cramer's Rule cannot be used to solve the system of equations.

The solution of the system is (x, y, z) where

$$x = \frac{D_x}{D}, \quad y = \frac{D_y}{D}, \quad z = \frac{D_z}{D}$$

Chapter 5
QUADRATIC EQUATIONS

IN THIS CHAPTER:

✔ *Quadratic Equations*
✔ *Methods of Solving Quadratic Equations*
✔ *Sum and Product of the Roots*
✔ *Nature of the Roots*
✔ *Radical Equations*
✔ *Systems of Equations Involving Quadratics*

Quadratic Equations

A quadratic equation in the variable x has the form $ax^2 + bx + c = 0$ where a, b, and c are constants and $a \neq 0$.

Thus, $x^2 - 6x + 5 = 0$, $2x^2 + x - 6 = 0$, and $3x^2 - 5 = 0$ are quadratic equations in one variable.

An incomplete quadratic equation is one which either b = 0 or c = 0, e.g., $4x^2 - 5 = 0$, $7x^2 - 2x = 0$, and $3x^2 = 0$.

To solve a quadratic equation is to find values of x which satisfy the equation. These values of x are called **zeros** or **roots** of the equation.

For example, $x^2 - 5x + 6 = 0$ is satisfied by x = 2 and x = 3. Then x = 2 and x = 3 are zeros or roots of the equation.

Methods of Solving Quadratic Equations

A.   **Solution by square root (when b = 0)**

Examples 5.1 Solve each quadratic equation for x.

(a) $x^2 - 4 = 0$ (b) $2x^2 - 21 = 0$ (c) $x^2 + 9 = 0$

(a) $x^2 - 4 = 0$. Then $x^2 = 4$, x = ± 2, and the roots are x = 2, - 2.

(b) $2x^2 - 21 = 0$. Then $x^2 = 21/2$ and the roots are

$$x = \pm \sqrt{\left(\frac{21}{2}\right)} = \pm \frac{1}{2}\sqrt{(42)}$$

(c) $x^2 + 9 = 0$. Then $x^2 = -9$ and the roots are $x = \pm \sqrt{(-9)} = \pm 3i$

B.   **Solution by factoring**

Examples 5.2 Solve each quadratic equation for x.

(a) $7x^2 - 5x = 0$ (b) $x^2 - 5x + 6 = 0$

(a) $7x^2 - 5x = 0$ may be written as $x(7x - 5) = 0$. Since the product of the two factors is zero, we set each factor equal to zero and solve the resulting linear equations. $x = 0$ or $7x - 5 = 0$. So $x = 0$ and $x = 5/7$ are the roots of the equation.

(b) $x^2 - 5x + 6 = 0$ may be written as $(x - 3)(x - 2) = 0$. Since the product is equal to 0, we set each factor equal to 0 and solve the resulting linear equations. $x - 3 = 0$ or $x - 2 = 0$. So $x = 3$ and $x = 2$ are the roots of the equation.

C. Solution by completing the square

Example 5.3 Solve $x^2 - 6x - 2 = 0$

Write the unknowns on one side and the constant term on the other; then

$$x^2 - 6x = 2$$

Add 9 to both sides, where 9 is the square of one-half of the coefficient of x, thus making the left-hand side a perfect square; then

$$x^2 - 6x + 9 = 2 + 9$$

or $(x - 3)^2 = 11$

Hence $x - 3 = \pm \sqrt{11}$ and the required roots are $x = 3 \pm \sqrt{11}$.

D. Solution by quadratic formula

The solutions of the quadratic equation $ax^2 + bx + c = 0$ are given by the formula

$$x = \frac{-b \pm \sqrt{b^2 - 4ac}}{2a}$$

where b^2 - 4ac is called the **discriminant** of the quadratic equation.

Example 5.4 Solve $3x^2$ - 5x + 1 = 0. Here a = 3, b = - 5, c = 1.

$$x = \frac{-(-5) \pm \sqrt{(-5)^2 - 4(3)(1)}}{2(3)} = \frac{5 \pm \sqrt{13}}{6}$$

Example 5.5 Solve $4x^2$ - 6x + 3 = 0. Here a = 4, b = - 6, c = 3.

$$x = \frac{-(-6) \pm \sqrt{(-6)^2 - 4(4)(3)}}{2(4)} = \frac{6 \pm \sqrt{-12}}{8} = \frac{6 \pm 2i\sqrt{3}}{8}$$

$$= \frac{2(3 \pm i\sqrt{3})}{8} = \frac{3 \pm i\sqrt{3}}{4}$$

E.   **Graphical solution**

The real roots or zeros of ax^2 + bx + c = 0 are the values of x corresponding to y = 0 on the graph of the parabola y = ax^2 + bx + c. Thus, the solutions are the abscissas of the points where the parabola intersects the x axis. If the graph does not intersect the x axis, the roots are imaginary.

Sum and Product of the Roots

The sum S and the product P of the roots of the quadratic equation ax^2 + bx + c = 0 are given by S = -b/a and P = c/a.

Thus, in $2x^2$ + 7x - 6 = 0, we have a = 2, b = 7, c = - 6 so that S = - 7/2 and P = - 6/2 = - 3.

It follows that a quadratic equation whose roots are r_1, r_2 is given by $x^2 - Sx + P = 0$ where $S = r_1 + r_2$ and $P = r_1 r_2$. Thus, the quadratic equation whose roots are $x = 2$ and $x = -5$ is $x^2 - (2 - 5)x + 2(-5) = 0$ or $x^2 + 3x - 10 = 0$.

Nature of the Roots

The nature of the roots of the quadratic equation $ax^2 + bx + c = 0$ is determined by the discriminant $b^2 - 4ac$. When the roots involve the imaginary unit i, we say the roots are imaginary.

Assuming a, b, c are *real numbers*, then

(1)   if $b^2 - 4ac > 0$, the roots are *real* and *unequal*,

(2)   if $b^2 - 4ac = 0$, the roots are *real* and *equal*,

(3)   if $b^2 - 4ac < 0$, the roots are *imaginary*.

Assuming a, b, c are *rational numbers*, then

(1)   if $b^2 - 4ac$ is a perfect square $\neq 0$, the roots are *real*, *rational*, and *unequal*,

(2)   if $b^2 - 4ac = 0$, the roots are *real*, *rational*, and *equal*,

(3)   if $b^2 - 4ac > 0$ but not a perfect square, the roots are *real*, *irrational*, and *unequal*,

(4)   if $b^2 - 4ac < 0$, the roots are *imaginary*.

Thus, $2x^2 + 7x - 6 = 0$, with discriminant $b^2 - 4ac = 7^2 - 4(2)(-6) = 97$, has roots which are real, irrational, and unequal.

Radical Equations

A radical equation is an equation having one or more unknowns under a radical. Thus,

$$\sqrt{x+3} - \sqrt{x} = 1 \quad \text{and} \quad \sqrt[3]{y} = \sqrt{y-4}$$

are radical equations.

To solve a radical equation, isolate one of the radical terms on one side of the equation and transpose all other terms to the other side. If both members of the equation are then raised to a power equal to the index of the isolated radical, the radical will be removed. This process is continued until radicals are no longer present.

Example 5.6 Solve

$$\sqrt{x+3} - \sqrt{x} = 1$$

Transposing, $\sqrt{x+3} = \sqrt{x} + 1$

Squaring $x+3 = x + 2\sqrt{x} + 1$ or $\sqrt{x} = 1$

Finally, squaring both sides gives $x = 1$.

Check:

$$\sqrt{1+3} - \sqrt{1} \ ? \ 1, \ 2-1 = 1$$

Systems of Equations Involving Quadratics

Graphical Solution

The real simultaneous solutions of two quadratic equations in x and y are the values of x and y corresponding to the points of intersection of the graphs of the two equations. If the graphs do not intersect, the simultaneous solutions are imaginary.

Algebraic Solution

A. **One linear and one quadratic equation**

Solve the linear equation for one of the unknowns and substitute in the quadratic equation.

Example 5.7 Solve the system

(1) $x + y = 7$

(2) $x^2 + y^2 = 25$

Solving (1) for y, $y = 7 - x$. Substitute in (2) and obtain $x^2 + (7 - x)^2 = 25$, $x^2 - 7x + 12 = 0$, $(x - 3)(x - 4) = 0$, and $x = 3, 4$. When $x = 3$, $y = 7 - x = 4$; when $x = 4$, $y = 7 - x = 3$. Thus, the simultaneous solutions are $(3, 4)$ and $(4, 3)$.

B. **Two equations of the form $ax^2 + by^2 = c$**

Use the method of addition or subtraction.

Example 5.8 Solve the system

(1) $2x^2 - y^2 = 7$

(2) $3x^2 + 2y^2 = 14$

To eliminate y, multiply (1) by 2 and add to (2); then

$7x^2 = 28,$ $x^2 = 4$ or $x = \pm 2$

Now put $x = 2$ or $x = -2$ in (1) and obtain $y = \pm 1$. The four solutions are: $(2, 1); (-2, 1); (2, -1); (-2, -1).$

C.   Two equations of the form $ax^2 + bxy + cy^2 = d$

Example 5.9 Solve the system

(1) $x^2 + xy = 6$

(2) $x^2 + 5xy - 4y^2 = 10$

Method 1.

Eliminate the constant term between both equations. Multiply (1) by 5, (2) by 3, and subtract; then

$x^2 - 5xy + 6y^2 = 0, (x - 2y)(x - 3y) = 0, x = 2y$ or $x = 3y.$

Now put $x = 2y$ in (1) or (2) and obtain $y^2 = 1, y = \pm 1.$

When $y = 1, x = 2y = 2$; when $y = -1, x = 2y = -2$. Thus, two solutions are: $x = 2, y = 1; x = -2, y = -1.$

Then put $x = 3y$ in (1) or (2) and get

$$y^2 = \frac{1}{2}, \quad y = \pm \frac{\sqrt{2}}{2}$$

When

$$y = \frac{\sqrt{2}}{2}, \quad x = 3y = \frac{3\sqrt{2}}{2}$$

When

$$y = -\frac{\sqrt{2}}{2}, \quad x = -\frac{3\sqrt{2}}{2}$$

Thus, the four solutions are:

$$(2,1); \quad (-2,-1); \quad \left(\frac{3\sqrt{2}}{2}, \frac{\sqrt{2}}{2} \right); \quad \left(-\frac{3\sqrt{2}}{2}, -\frac{\sqrt{2}}{2} \right)$$

Method 2.

Let y = mx in both equations.

From (1):

$$x^2 + mx^2 = 6, \quad x^2 = \frac{6}{1+m}$$

From (2):

$$x^2 + 5mx^2 - 4m^2x^2 = 10, \quad x^2 = \frac{10}{1+5m-4m^2}$$

Then

$$\frac{6}{1+m} = \frac{10}{1+5m-4m^2}$$

from which m = 1/2, 1/3; hence y = x/2, y = x/3. The solution proceeds as in Method 1.

Chapter 6
SEQUENCES, SERIES, AND MATHEMATICAL INDUCTION

IN THIS CHAPTER:

✔ *Sequences*
✔ *Arithmetic Sequences*
✔ *Geometric Sequences*
✔ *Infinite Geometric Series*
✔ *Harmonic Sequences*
✔ *Means*
✔ *Mathematical Induction*

Sequences

A **sequence** of numbers is a function defined on the set of positive integers. The numbers in the sequence are called **terms**. A **series** is the sum of terms of a sequence.

Arithmetic Sequences

An **arithmetic sequence** is a sequence of numbers each of which, after the first, is obtained by adding to the preceding number a constant number called the **common difference**.

Thus 3, 7, 11, 15, 19, ... is an arithmetic sequence because each term is obtained by adding 4 to the preceding number. In the arithmetic sequence 50, 45, 40, ... the common difference is 45 - 50 = 40 - 45 = -5.

General formulas for arithmetic sequences include:

- The nth term, or last term: $l = a + (n - 1)d$

- The sum of the first n terms:

$$S = \frac{n}{2}(a+l) = \frac{n}{2}[2a+(n-1)d]$$

where a = first term of the sequence;
d = common difference;
n = number of terms;
l = nth term, or last term; and,
S = sum of first n terms.

Example 6.1 Consider the arithmetic sequence 3, 7, 11, ... where a = 3 and d = 7 - 3 = 11 - 7 = 4. The sixth term is $l = a + (n - 1)d = 3 + (6 - 1)4 = 23$. The sum of the first six terms is:

$$S = \frac{n}{2}[2a+(n-1)d] = \frac{6}{2}[2(3)+(6-1)4] = 78$$

or

$$S = \frac{n}{2}(a+l) = \frac{6}{2}[3+23] = 78$$

Geometric Sequences

A **geometric sequence** is a sequence of numbers each of which, after the first, is obtained by multiplying the preceding number by a constant number called the **common ratio**.

Thus, 5, 10, 20, 40, 80, ... is a geometric sequence because each number is obtained by multiplying the preceding number by 2. In the geometric sequence 9, -3, 1, -1/3, 1/9, ... the common ratio is

$$\frac{-3}{9} = \frac{1}{-3} = \frac{-1/3}{1} = \frac{1/9}{-1/3} = -\frac{1}{3}$$

General formulas for arithmetic sequences include:

- The nth term, or last term: $l = ar^{n-1}$

- The sum of the first n terms:

$$S = \frac{a(r^n - 1)}{r-1} = \frac{rl-a}{r-1}, \ r \neq 1$$

where   a = first term;
        r = common ratio;
        n = number of terms;
        l = nth term, or last term; and,
        S = sum of first n terms.

Example 6.2 Consider the geometric sequence 5, 10, 20, ... where a = 5 and

$$r = \frac{10}{5} = \frac{20}{10} = 2$$

The seventh term is $l = ar^{n-1} = 5(2^{7-1}) = 5(2^6) = 320$. The sum of the first seven terms is:

$$S = \frac{a(r^n - 1)}{r - 1} = \frac{5(2^7 - 1)}{2 - 1} = 635$$

Infinite Geometric Series

The sum to infinity (S_∞) of any geometric sequence in which the common ratio r is numerically less than 1 is given by

$$S_\infty = \frac{a}{1 - r}, \quad \text{where } |r| < 1$$

Example 6.3 Consider the infinite geometric series

$$1 - \frac{1}{2} + \frac{1}{4} - \frac{1}{8} + \cdots$$

where a = 1 and r = -1/2. Its sum to infinity is

$$S_\infty = \frac{a}{1 - r} = \frac{1}{1 - (-1/2)} = \frac{1}{3/2} = \frac{2}{3}$$

Harmonic Sequences

A **harmonic sequence** is a sequence of numbers whose reciprocals form an arithmetic sequence.

Thus,

$$\frac{1}{2}, \frac{1}{4}, \frac{1}{6}, \frac{1}{8}, \frac{1}{10}, \cdots$$

is a harmonic sequence because 2, 4, 6, 8, 10, ... is an arithmetic sequence.

Example 6.4 Compute the 15th term of the harmonic sequence

$$\frac{1}{4}, \frac{1}{7}, \frac{1}{10}, \cdots$$

The corresponding arithmetic sequence is 4, 7, 10, ...; its 15th term is l = a + (n - 1)d = 4 + (15 - 1)3 = 46. Hence the 15th term of the harmonic progression is 1/46.

Means

The terms between any two given terms of sequence are called the **means** between these two terms.

Thus, in the arithmetic sequence 3, 5, 7, 9, 11, ... the arithmetic mean between 3 and 7 is 5, and *four* arithmetic means between 3 and 13 are 5, 7, 9, 11.

In the geometric sequence 2, -4, 8, -16, ... *two* geometric means between 2 and -16 are -4, 8.

In the harmonic sequence

$$\frac{1}{2}, \frac{1}{3}, \frac{1}{4}, \frac{1}{5}, \frac{1}{6}, \cdots$$

the harmonic mean between 1/2 and 1/4 is 1/3, and *three* harmonic means between 1/2 and 1/6 are 1/3, 1/4, 1/5.

Example 6.5 What is the harmonic mean between 3/8 and 4?

The arithmetic mean between 8/3 and 1/4 is

$$\frac{1}{2}\left(\frac{8}{3} + \frac{1}{4}\right) = \frac{35}{24}$$

where the arithmetic mean between two terms A and B is always $\frac{A+B}{2}$. Hence the harmonic mean between 3/8 and 4 = 24/35.

Mathematical Induction

Principle of Mathematical Induction

Some statements are defined on the set of positive integers. To establish the truth of such a statement, we could prove it for each positive integer of interest separately. However, since there are infinitely many positive integers, this case-by-case procedure can never prove that the statement is always true. A procedure called **mathematical induction** can be used to establish the truth of the statement for all positive integers.

Principle of Mathematical Induction

Let P(n) be a statement that is either true or false for each positive integer n. P(n) is true for all positive integers, n, if the following two conditions are satisfied:

(1)　　P(1) is true.

(2)　　Whenever n = kP(k) is true, it is implied that P(k + 1) is true.

Proof by Mathematical Induction

To prove a theorem or formula by mathematical induction, there are two distinct steps in the proof:

(1)　　Show by actual substitution that the proposed theorem or formula is true for some one positive integer n, such as n = 1, or n = 2, etc.

(2)　　Assume that the theorem or formula is true for n = k. Then prove that it is true for n = k + 1.

Once both steps have been completed, then you can conclude that the theorem or formula is true for all positive integers greater than or equal to a, the positive integer from the first step.

Example 6.6 Prove by mathematical induction that, for all positive integers n,

$$1+2+3+ \ldots +n=\frac{n(n+1)}{2}$$

Step 1. The formula is true for n = 1, since

$$1 = \frac{1(1+1)}{2} = 1$$

Step 2. Assume that the formula is true for n = k. Then, adding (k + 1) to both sides,

$$1 + 2 + 3 + \dots + k + (k+1) = \frac{k(k+1)}{2} + (k+1) = \frac{(k+1)(k+2)}{2}$$

which is the value of n(n + 1)/2 when (k + 1) is substituted for n.

Hence, if the formula is true for n = k, we have proved it to be true for n = k + 1. But the formula holds for n = 1; hence it holds for n = 1 + 1 = 2. Then, since it holds for n = 2, it holds for n = 2 + 1 = 3, and so on. Thus, the formula is true for all positive integers n.

Chapter 7
PERMUTATIONS, COMBINATIONS, THE BINOMIAL THEOREM, AND PROBABILITY

IN THIS CHAPTER:

✔ *Fundamental Counting Principle*
✔ *Permutations*
✔ *Combinations*
✔ *Combinatorial Notation*
✔ *The Binomial Theorem*
✔ *Simple Probability*
✔ *Compound Probability*
✔ *Binomial Probability*
✔ *Conditional Probability*

Fundamental Counting Principle

If one thing can be done in m different ways and, when it is done in any one of these ways, a second thing can be done in n different ways, then the two things in succession can be done in mn different ways.

For example, if there are 3 candidates for governor and 5 for mayor, then the two offices may be filled in $3 \cdot 5 = 15$ ways.

In general, if a_1 can be done in x_1 ways, a_2 can be done in x_2 ways, a_3 can be done in x_3 ways, and a_n can be done in x_n ways, then the event $a_1 a_2 a_3 \cdots a_n$ can be done in $x_1 \cdot x_2 \cdot x_3 \cdots x_n$ ways.

Example 7.1 A man has 3 jackets, 10 shirts, and 5 pairs of slacks. If an outfit consists of a jacket, a shirt, and a pair of slacks, how many different outfits can the man make?

$x_1 \cdot x_2 \cdot x_3 = 3 \cdot 10 \cdot 5 = 150$ outfits

Permutations

A **permutation** is an arrangement of all or part of a number of things in a definite order.

For example, the permutations of the three letters a, b, c taken all at a time are abc, acb, bca, bac, cba, cab. The permutations of the three letters a, b, c taken two at a time are ab, ac, ba, bc, ca, cb.

For a natural number n, n factorial, denoted by n!, is the product of the first n natural numbers. That is, $n! = n \cdot (n - 1) \cdot (n - 2) \cdots 2 \cdot 1$. Also, $n! = n \cdot (n - 1)!$. Zero factorial is defined to be 1 or $0! = 1$.

Examples 7.2 Evaluate each factorial.

(a) 7!   (b) 5!   (c) 1!   (d) 2!   (e) 4!

(a) $7! = 7 \cdot 6 \cdot 5 \cdot 4 \cdot 3 \cdot 2 \cdot 1 = 5040$
(b) $5! = 5 \cdot 4 \cdot 3 \cdot 2 \cdot 1 = 120$
(c) $1! = 1$
(d) $2! = 2 \cdot 1 = 2$
(e) $4! = 4 \cdot 3 \cdot 2 \cdot 1 = 24$

The symbol $_nP_r$ represents the number of permutations (arrangements, orders) of n things taken r at a time. Thus, $_8P_3$ denotes the number of permutations of 8 things taken 3 at a time, and $_5P_5$ denotes the number of permutations of 5 things taken 5 at a time.

 **Note!**

The symbol P(n, r) having the same meaning as $_nP_r$ is sometimes used.

Permutations of n Different Things Taken r at a Time

$$_nP_r = n(n-1)(n-2)\cdots(n-r+1) = \frac{n!}{(n-r)!}$$

when r = n, $_nP_r = {_nP_n} = n(n-1)(n-2)\cdots 1 = n!$.

Examples 7.3 Evaluate the following permutations:

(a) $_5P_1$ (b) $_5P_2$ (c) $_5P_3$ (d) $_5P_4$ (e) $_5P_5$

(a) $_5P_1 = 5$

(b) $_5P_2 = 5 \cdot 4 = 20$

(c) $_5P_3 = 5 \cdot 4 \cdot 3 = 60$

(d) $_5P_4 = 5 \cdot 4 \cdot 3 \cdot 2 = 120$

(e) $_5P_5 = 5 \cdot 4 \cdot 3 \cdot 2 \cdot 1 = 120$

Example 7.4 Determine the number of ways in which 4 persons can take their places in a cab having 6 seats.

$_6P_4 = 6 \cdot 5 \cdot 4 \cdot 3 = 360$

Permutations with Some Things Alike, Taken All at a Time

The number of permutations P of n things taken all at a time, of which n_1 are alike, n_2 others are alike, n_3 others are alike, etc., is

$$P = \frac{n!}{n_1!n_2!n_3! \cdots}$$

where $n_1 + n_2 + n_3 + \cdots = n$.

Example 7.5 The number of ways 3 dimes and 7 quarters can be distributed among 10 boys, each to receive one coin, is

$$\frac{10!}{3!7!} = \frac{10 \cdot 9 \cdot 8}{1 \cdot 2 \cdot 3} = 120$$

Circular Permutations

The number of ways of arranging n different objects around a circle is (n - 1)! ways.

Example 7.6 Ten persons may be seated at a round table in $(10 - 1)! = 9!$ ways.

Combinations

A **combination** is a grouping or selection of all or part of a number of things without reference to the arrangement of the things selected.

Thus, the combinations of the three letters a, b, c taken 2 at a time are ab, ac, bc. Note that ab and ba are 1 combination but 2 permutations of the letters a and b.

The symbol $_nC_r$ represents the number of combinations (selections, groups) of n things taken r at a time.

Thus, $_9C_4$ denotes the number of combinations of 9 things taken 4 at a time.

 **Note!**

The symbol C(n, r) having the same meaning as $_nC_r$ is sometimes used.

Combinations of n Different Things Taken r at a Time

$$_nC_r = \frac{_nP_r}{r!} = \frac{n!}{r!\,(n-r)!} = \frac{n\,(n-1)\,(n-2)\cdots(n-r+1)}{r!}$$

Example 7.7 The number of handshakes that may be exchanged among a party of 12 students if each student shakes hands once with each other student is

$$_{12}C_2 = \frac{12!}{2!(12-2)!} = \frac{12!}{2!10!} = \frac{12\cdot11}{1\cdot2} = 66$$

The following formula is very useful in simplifying calculations:

$$_nC_r = {_nC_{n-r}}$$

This formula indicates that the number of selections of r out of n things is the same as the number of selections of n - r out of n things.

Example 7.8 Evaluate the following combinations:

(a) $_5C_1$ (b) $_5C_2$ (c) $_5C_3$ (d) $_5C_4$ (e) $_5C_5$

(a) $_5C_1 = \frac{5}{1} = 5;$ (b) $_5C_2 = \frac{5\cdot4}{1\cdot2} = 10;$ (c) $_5C_3 = \frac{5\cdot4\cdot3}{1\cdot2\cdot3} = 10;$

(d) $_5C_4 = \frac{5\cdot4\cdot3\cdot2}{1\cdot2\cdot3\cdot4} = 5;$ (e) $_5C_5 = \frac{5\cdot4\cdot3\cdot2\cdot1}{1\cdot2\cdot3\cdot4\cdot5} = 1$

Note that, in each case, the numerator and denominator have the same number of factors.

Combinations of Different Things Taken Any Number at a Time

The total number of combinations C of n different things taken 1, 2, 3, ..., n at a time is

$C = 2^n - 1$

Example 7.9 A woman has in her pocket a quarter, a dime, a nickel, and a penny. The total number of ways she can draw a sum of money from her pocket is $2^4 - 1 = 15$.

Combinatorial Notation

The number of combinations of n objects selected r at a time, $_nC_r$, can be written in the form

$$\binom{n}{r}$$

which is called combinatorial notation.

$$C_r = \frac{n!}{(n-r)!r!} = \binom{n}{r}$$

where n and r are integers and $r \le n$.

Examples 7.10 Evaluate each expression.

(a) $\binom{7}{3}$ (b) $\binom{8}{7}$ (c) $\binom{9}{9}$ (d) $\binom{5}{0}$

(a) $\binom{7}{3} = \frac{7!}{(7-3)!3!} = \frac{7!}{4!3!} = \frac{7 \cdot 6 \cdot 5 \cdot 4!}{4!3 \cdot 2 \cdot 1} = 7 \cdot 5 = 35$

(b) $\binom{8}{7} = \frac{8!}{(8-7)!7!} = \frac{8!}{1!7!} = \frac{8 \cdot 7!}{1 \cdot 7!} = 8$

(c) $\binom{9}{9} = \frac{9!}{(9-9)!9!} = \frac{9!}{0!9!} = \frac{1}{0!} = \frac{1}{1} = 1$

(d) $\binom{5}{0} = \frac{5!}{(5-0)!0!} = \frac{5!}{5!0!} = \frac{1}{0!} = \frac{1}{1} = 1$

The Binomial Theorem

If n is a positive integer, we expand $(a + x)^n$ as shown below:

$$(a+x)^n = a^n + na^{n-1}x + \frac{n(n-1)}{2!}a^{n-2}x^2 + \frac{n(n-1)(n-2)}{3!}a^{n-3}x^3$$

$$+ \cdots + \frac{n(n-1)(n-2)\cdots(n-r+2)}{(r-1)!}a^{n-r+1}x^{r-1} + \cdots + x^n$$

This equation is called the **binomial theorem,** or binomial formula.

Other forms of the binomial theorem exist and some use combinations to express the coefficients. The relationship between the coefficients and combinations are shown below.

$$\frac{5 \cdot 4}{2!} = \frac{5 \cdot 4}{2!}\left(\frac{3 \cdot 2 \cdot 1}{3 \cdot 2 \cdot 1}\right) = \frac{5 \cdot 4 \cdot 3 \cdot 2 \cdot 1}{3 \cdot 2 \cdot 1 \cdot 2!} = \frac{5!}{3!2!} = \frac{5!}{(5-2)!2!} = \binom{5}{2}$$

$$\frac{n(n-1)(n-2)}{3!} = \frac{n(n-1)(n-2)\cdots 2 \cdot 1}{(n-3)3!} = \frac{n!}{(n-3)3!} = \binom{n}{3}$$

So

$$(a+x)^n = a^n + \frac{n!}{(n-1)!1!}a^{n-1}x + \frac{n!}{(n-2)!2!}a^{n-2}x^2 + \cdots$$

$$+ \frac{n!}{(n-[r-1])!(r-1)!}a^{n-r+1}x^{r-1} + \cdots + x^n$$

and

$$(a+x)^n = a^n + \binom{n}{1}a^{n-1}x + \binom{n}{2}a^{n-2}x^2 + \cdots$$

$$+ \binom{n}{r-1}a^{n-r+1}x^{r-1} + \cdots + x^n$$

Note that in the expansion of $(a + x)^n$:

(1) The exponent of a + the exponent of x = n (i.e., the degree of each term is n).

(2) The number of terms is n + 1, where n is a positive integer.

(3) There are **two** middle terms when n is an **odd** positive integer.

(4) There is only **one** middle term when n is an **even** positive integer.

(5) The coefficients of the terms which are equidistant from the ends are the same. It is interesting to note that these coefficients may be arranged as follows:

$(a + x)^0$					1					
$(a + x)^1$				1		1				
$(a + x)^2$			1		2		1			
$(a + x)^3$		1		3		3		1		
$(a + x)^4$	1		4		6		4		1	
$(a + x)^5$	1	5		10		10		5		1

This array of numbers is known as **Pascal's Triangle**. The first and last numbers in each row are 1, while any other number in the array can be obtained by adding the two numbers to the right and left of it in the preceding row.

Example 7.11 Expand $(a + x)^3$.

$$(a+x)^3 = a^3 + 3a^2x + \frac{3\cdot 2}{1\cdot 2}ax^2 + \frac{3\cdot 2\cdot 1}{1\cdot 2\cdot 3}x^3 = a^3 + 3a^2x + 3ax^2 + x^3$$

The rth term formula for the expression of $(a + x)^n$ can be expressed in terms of combinations.

$$\text{rth term} = \frac{n(n-1)(n-2)\cdots(n-r+2)}{(r-1)!}a^{n-r+1}x^{r-1}$$

$$= \frac{n(n-1)(n-2)\cdots(n-r+2)(n-r+1)\cdots 2\cdot 1}{(n-r+1)(n-r)\cdots 2\cdot 1(r-1)!}a^{n-r+1}x^{r-1}$$

$$\text{rth term} = \frac{n!}{(n-[r-1])!(r-1)!}a^{n-r+1}x^{r-1}$$

$$\text{rth term} = \binom{n}{r-1}a^{n-r+1}x^{r-1}$$

Example 7.12 Compute the sixth term of $(x + y)^{15}$ using the formula

$$\text{rth term of }(a+x)^n = \frac{n(n-1)(n-2)\cdots(n-r+2)}{(r-1)!}a^{n-r+1}x^{r-1}$$

In this case, n = 15, r = 6, n - r + 2 = 11, r - 1 = 5, n - r + 1 = 10.

$$\text{6th term} = \frac{15\cdot 14\cdot 13\cdot 12\cdot 11}{1\cdot 2\cdot 3\cdot 4\cdot 5}x^{10}y^5 = 3003x^{10}y^5$$

Simple Probability

Suppose that an event can happen in h ways and fail to happen in f ways, all these h + f ways supposed equally likely. Then the probability of the occurrence of the event (called its success) is

$$p=\frac{h}{h+f}=\frac{h}{n}$$

and the probability of the non-occurrence of the event (called its failure) is

$$q=\frac{f}{h+f}=\frac{f}{n}$$

where n = h + f. It follows that:

p + q = 1, p = 1 - q, and q = 1 - p.

The odds in favor of the occurrence of the event are h:f or h/f; the odds against its happening are f:h or f/h. If p is the probability that an event will occur, the odds in favor of its happening are p:q = p:(1 - p) or p/(1 - p); the odds against its happening are q:p = (1 - p):p or (1 - p)/p.

Compound Probability

Two or more events are said to be independent if the occurrence or non-occurrence of any one of them does not affect the probabilities of occurrence of any of the others.

Thus, if a coin is tossed four times and it turns up a head each time, the fifth toss may be head or tail and is not influenced by the previous tosses.

The probability that two or more independent events will happen is equal to the product of their separate probabilities.

Thus, the probability of getting a head on both the fifth and sixth tosses is $1/2(1/2) = 1/4$.

Two or more events are said to be dependent if the occurrence or non-occurrence of one of the events affects the probabilities of occurrence of any of the others.

Consider that two or more events are dependent. If p_1 is the probability of a first event, p_2 the probability that after the first event has happened the second will occur, p_3 the probability that after the first and second events have happened the third event will occur, etc., then the probability that all events will happen in the given order is the product $p_1 \cdot p_2 \cdot p_3 \cdot \cdot \cdot$.

Two or more events are said to be mutually exclusive if the occurrence of any one of them excludes the occurrence of the others.

The probability of occurrence of one of two or more mutually exclusive events is the *sum* of the probabilities of the individual events.

Example 7.13 If a die is thrown, what is the probability of getting a 5 or a 6?

Getting a 5 and getting a 6 are mutually exclusive so

$$P\left(5 \text{ or } 6\right) = P\left(5\right) + P\left(6\right) = \frac{1}{6} + \frac{1}{6} = \frac{2}{6} = \frac{1}{3}$$

Two events are said to be overlapping if the events have at least one outcome in common, hence they can happen at the same time. The probability of occurrence of one of two overlapping events is the sum of probabilities of the two individual events minus the probability of their common outcomes.

Example 7.14 If a die is thrown, what is the probability of getting a number less than 4 or an even number?

The numbers less than 4 on a die are 1, 2, and 3. The even numbers on a die are 2, 4, and 6. Since these two events have a common outcome, 2, they are overlapping events.

P(less than 4 or even) =
$$P(\text{less than 4}) + P(\text{even}) - P(\text{less than 4 and even})$$

$$= \frac{3}{6} + \frac{3}{6} - \frac{1}{6} = \frac{5}{6}$$

Binomial Probability

If p is the probability that an event will happen in any single trial and q = 1 - p is the probability that it will fail to happen in any single trial, then the probability of its happening exactly r times in n trials is $_nC_rp^rq^{n-r}$. The probability that an event will happen at least r times in n trials is:

$$p^n + {}_nC_1p^{n-1}q + {}_nC_2p^{n-2}q^2 + \bullet \bullet \bullet + {}_nC_rp^rq^{n-r}$$

This expression is the sum of the first n - r + 1 terms of the binomial expansion of $(p + q)^n$.

Conditional Probability

The probability that a second event will occur given that the first event has occurred is called conditional probability. To find the probability that the second event will occur given that the first event occurred, divide the probability that both events occurred by the probability of the first event. The probability of event B given that event A has occurred is denoted by P(B|A).

Example 7.15 A box contains black chips and red chips. A person draws two chips without replacement. If the probability of selecting a black chip and a red chip is 15/56 and the probability of drawing a black chip on the first draw is 3/4, what is the probability of drawing a red chip on the second draw, if you know the first chip drawn was black?

If B is the event drawing a black chip and R is the event drawing a red chip, then P(R|B) is the probability of drawing a red chip on the second draw given that a black chip was drawn on the first draw.

$$P(R|B) = \frac{P(R \text{ and } B)}{P(B)} = \frac{15/56}{3/4} = \frac{15}{56} \cdot \frac{4}{3} = \frac{5}{14}$$

Thus, the probability of drawing a red chip on the second draw given that a black chip was drawn on the first draw is 5/14.

Index

Abscissa, 45
Absolute inequality, 87
Addition, 2, 5–7, 21, 31, 81, 106
Algebraic expressions, 19–25
Algebraic fractions, 33–37
Arithmetic mean, 114–15
Arithmetic sequence, 111–12
Associative properties, 22
Asymptotes, 61–64

Binomial, 19, 25, 27, 125, 127, 130
Braces, 21
Brackets, 21

Circle, 72, 121
Coefficients, 23, 29–30, 32–33, 50, 52–56, 58, 67, 69, 81, 97, 99, 125–26
Combinations, 32, 122–25, 127
Common difference, 111
Common ratio, 112–13
Completing the square, 102
Complex fractions, 36–37
Complex numbers, 15–17, 50
Complex roots, 55
Compound probability, 128–30
Conditional equation, 70–71
Conditional inequality, 87
Conditional probability, 130–31
Constant, 12, 20, 40–41, 48, 50, 52–53, 61, 63, 66, 77, 102, 107, 111–12

Continuity, 56
Coordinate system, 45
Coordinates, 6, 15, 42, 45, 82
Counting principle, 119
Cramer's Rule, 96–98
Cube of a binomial, 27
Cubic equation, 74

Decimal, 59–60
Decomposition, 68–69
Degree, 20, 24, 32–33, 50–52, 54–56, 59, 62–63, 65, 72–74, 89, 126
Denominator, 2, 8–9, 14, 17, 33–37, 62–63, 65–67, 91–92, 97, 123
Dependent equations, 83
Dependent variable, 43–44
Descartes' Rule of Signs, 58–59
Determinants, 94–99
Difference, 2, 7, 9, 11, 25, 30–31, 111
Discriminant, 103–04
Dividend, 2, 24, 53–54
Division, 2, 52–53, 56–57, 59–60, 65, 71
Divisor, 2, 24, 36, 53–54
Domain, 42–44, 61–63
Double root, 55

Equations, 39, 42–43, 47, 50, 52, 55, 70–109

Equivalent fractions, 35
Exponential form, 11–12
Exponents, 9–10, 20, 22, 50, 72
Extremes, 39

Factor, 18–20, 29–30, 32–33, 52, 54–56, 61, 66–67, 90, 92, 102
Factor theorem, 52
Factoring, 29–32, 101
Formulas, 72, 111–12
Fourth proportional, 39
Fractions, 2, 8–9, 14, 33–37, 64–67, 74
Function, 41–44, 50–52, 61–64
Fundamental Counting Principle, 119
Fundamental Theorem of Algebra, 54
Fundamental theorems, 66–68
Fundamental operations, 1–5

Geometric means, 114
Geometric sequence, 112–13
Geometric series, 113
Graphs, 41, 46–50, 63–64, 106
Graphical representation, 3–4, 15–16
Greatest common factor, 32–33
Grouping, 21, 31, 122

Harmonic sequence, 114
Horizontal asymptoics, 62–63

Identity, 4–6, 70–71
Imaginary unit, 104
Improper fraction, 65
Independent events, 128
Independent variable, 43
Index, 12, 14, 105

Induction, 115–17
Inequalities, 86–94
Infinite geometric series, 113
Infinity, 113
Integers, 2–3, 6–7, 10, 30, 56, 72, 110, 115–17, 124
Integral root theorem, 56
Interception form, 80
Interest, 115
Intermediate Value Theorem, 56–57, 59–60
Inverse property, 5
Irrational number, 12
Irrational roots, 55
Inverse property, 5

Least common multiple, 33
Like terms, 20–21, 74
Linear equations, 70–99, 102
Lines, 45, 63, 76–80, 90–92
Literals, 20
Logarithms, 10–12

Mathematical induction, 110–11, 113, 115–17
Mean proportional, 39
Means, 114–15
Minuend, 21
Monomial, 19–20, 22–23, 25, 30, 72
Monomial factor, 30
Multinomial, 19–20
Multiplication, 2, 5–6, 22, 28
Mutually exclusive events, 129

Natural logarithms, 12
Natural numbers, 2, 119
Notation, 44, 124
Number system, 3–4

Numbers, 1–8, 15–17, 19–20, 30, 38, 42–44, 50, 53, 57, 59–60, 77, 81, 84, 89, 95, 98, 104, 110–12, 114, 119, 126, 130
Numerator, 2, 8–9, 17, 33–37, 44, 62–63, 65, 91–92, 97, 123

Odds, 128
Operations, 1–5, 8–9, 16–17, 35–36, 71–72
Ordinate, 45
Origin, 3–4, 7, 42, 45, 47

Parabola, 103
Parentheses, 2, 21, 74, 91
Partial fractions, 65, 68–69
Pascal's Triangle, 126–27
Perfect nth powers, 13–14
Permutations, 119–22
Point, 3–4, 7, 14–15, 42–43, 45, 63, 77–78, 82–84, 86, 92–93
Polynomial equations, 72–74
Polynomial functions, 50–60
Polynomials, 20, 23–24, 29, 32–33, 61, 64, 69
Positive numbers, 11
Powers, 1, 9, 13–14, 20, 23–24, 53, 58, 72
Principal, 12
Probability, 118, 128–31
Product, 2, 5–6, 9, 11, 25–29, 32–33, 36, 53, 89–91, 100, 102–03, 119, 128–29
Products, 6, 25–29
Proper fraction, 65
Properties of numbers, 5–9
Proportion, 39–40
Proportional, 39, 41
Proportionality, 40

Quadrants, 45
Quadratic equations, 100–10
Quadratic formula, 102
Quotient, 2, 11, 23–24, 33, 36, 52–54, 58, 64, 89, 91

Radical equations, 105
Radicals, 12–14, 105
Radicand, 12–14
Ratio, 38–40, 61, 77, 112–13
Rational fractions, 64
Rational function, 61–64
Rational number, 2, 4
Rational root theorem, 55
Real numbers, 2–9, 15, 30, 42–44, 77, 84, 89, 104
Reciprocal, 9, 72
Rectangular coordinate system, 45
Relation, 42, 47, 72
Remainder, 24, 52, 54
Remainder theorem, 52
Roots, 32, 52, 54–59, 71–72, 101, 103–04
Rules of signs, 7–8

Scaling, 49–50
Sense of an inequality, 87
Sequence, 110–13
Series, 110–11, 113, 115, 117
Sets, 4–5
Shifts, 48–49
Signs, 7–8, 22, 34, 56–59, 81, 86, 90–92
Simple probability, 128
Simultaneous linear equations, 81–84, 96
Slope, 76–78

Solutions, 71, 83, 92, 102–03, 106–08
Special products, 25–28
Square, 25, 27, 30, 32, 40–41, 43, 87, 101–02, 104
Subtraction, 2, 21–22, 31, 81, 106
Subtrahend, 21
Sum, 2, 5–6, 8–9, 11, 20, 25, 31, 35, 53–54, 65–66, 72, 76, 100, 103, 110–13, 123, 129–30
Symmetry, 45–47, 63
Synthetic division, 52–53, 56–57, 59–60

Systems of equations, 84–86, 106–09

Terms, 19–20, 50, 53–54, 62, 72, 74, 87, 102, 107, 111–14, 126–27
Trinomial, 19, 27

Variable, 41, 43–44, 50–51, 53, 72, 74–75, 97, 100
Variation, 40–41
Vertical asymptotes, 61

Zero, 2–3, 8, 20, 24, 34, 43–44, 53, 56–60, 63, 67, 71–72, 77, 81, 84, 87–89, 92, 99, 102, 119